A MOUNTAIN TO CLIMB

HAKAN BULGURLU

This edition first published in English in 2022
In partnership with whitefox publishing

Hardback ISBN: 978-1-913532-92-5
eBook ISBN: 978-1-913532-93-2

To find out more about the author and book, please visit amountaintoclimb.org

Cover design by Salon Couture Books
Designed and typeset by seagulls.net

Printed and bound in Turkey by PrintCenter.com.tr, Istanbul

To Sasha, Andrea and Oscar;

and all the children of the pale blue dot.

Sevgili Desha,

Keyifle okuman dileklerimle.

Hakan Oyfle

CONTENTS

'Look again at that dot. That's here. That's home. That's us. On it everyone you love, everyone you know, everyone you ever heard of, every human being who ever was, lived out their lives ... every mother and father, hopeful child, inventor and explorer, every teacher of morals, every corrupt politician, every "superstar," every "supreme leader," every saint and sinner in the history of our species lived there – on a mote of dust suspended in a sunbeam.'

– Carl Sagan, *Pale Blue Dot*

Earthrise by Bill Anders. (Source: NASA)

INTRODUCTION

On Christmas Eve, 1968, Bill Anders was one of three astronauts on board the Apollo 8 spacecraft when he saw something out of the hatch window. The mission was the first manned flight to orbit the moon. The craft was on its fourth orbit, emerging from the dark side of the moon, when Anders saw the blue and white of earth emerging out of the deep black of space. 'Oh my God!' Anders exclaimed. 'Look at that picture over there! Here's the earth coming up. Wow, is that pretty!' Anders reached for his Hasselblad camera to try and capture the moment. The first picture he took was in black and white. Then one of his fellow astronauts, Jim Lovell, found him a roll of colour film. Anders took the shot that would become known as *Earthrise* – the first colour photo of our planet taken from space.

More than fifty years on, *Earthrise* remains one of the most iconic pictures of all time, a shot described by nature photographer Galen Rowell as 'the most influential environmental picture ever taken.' 'To see the earth as it truly is,' the poet Archibald MacLeish wrote in the *New York Times*, 'small and blue and beautiful in that eternal silence where it floats, is to see ourselves as riders on the earth together, brothers on that bright loveliness in the eternal cold.'

But for all the undimmed beauty and power of Anders's photograph, the self-awareness and understanding of who we are has sadly not gone on to be reflected in our stewardship of this solitary jewel. Over the five decades since that photograph was taken, the damage done to the planet has accelerated to the point that scientists now talk of the Anthropocene, a new era in the planet's history where, for the first time, human activity is having a significant impact on the planet's climate and ecosystems.

The world is heating up at a rate that is unsustainable. The signs of climate change are all around us, and yet human behaviour is barely altered in response. Our planet is under threat like never before. There's evidence of this everywhere, from the bleaching of coral reefs to the melting of glaciers, and in extreme weather events like mass flooding and devastating forest fires. Our oceans are scarred with plastic waste we can see, and microplastics and fibres we can't. Biodiversity continues to decline. Efforts like COP26, the UN's Climate Change Conference, are valuable, but the situation continues to be dire, and we are fast approaching a global tipping point. The clock is ticking.

* * *

One of the challenges in talking about the environment is that it is easy for the eyes to glaze over. The problem can feel so large and overwhelming, it's difficult to know where to start. The easier option is to change the channel and hope that someone else will sort things out. But as Patriarch Bartholomew, the spiritual leader of Eastern Orthodox Christians worldwide, puts it, 'Today there are no excuses for our lack of involvement. If we do not choose to care, then we are not simply indifferent onlookers; we are in fact active aggressors.' If we want to look our children and grandchildren in the eye, then we all need to step up: as leaders, as business people, as individuals.

As a business leader myself, I've been on a journey and a half over the last few years to do what I can to make my company more environmentally responsible. I'm the CEO of Arçelik, one of the largest white goods manufacturers in the world. We have a dozen household brands that include Beko and Grundig, and have a workforce of over 43,000 people. We have sales and marketing offices in more than forty countries and twenty-eight production facilities across nine. In total, Arçelik offers products and services in nearly 150 countries.

Under my leadership, Arçelik has continued its long-standing goal of sustainability. For three years in a row (2019, 2020, 2021), the company was the highest scored home appliances company on the Dow Jones Sustainability index – an index which evaluates the sustainability performance of the world's largest companies.

In 2020, Arçelik became carbon neutral in its global production. For a company our size, in the industry we're in, and in the country where our headquarters is based, this is no small achievement. By introducing innovative refrigerators with high energy efficiency levels, the company was able to obtain sufficient carbon credits to offset the direct and indirect greenhouse gas emissions generated in our global production facilities. It's a fantastic achievement for the organisation and a testament to the hard work and creativity of our team. Arçelik is the R&D leader in Turkey – we have over 1,600 researchers across the company, and from solar-panelled fridges to microfibre filters on our washing machines, we are constantly looking for ways to make our products better for the planet. I'm not telling you this to boast, but to show you that whatever industry you work in, change is possible. It's not always easy, but it can be done in a way that is both sustainable for the planet and also for the company's balance sheet.

When I took over at Arçelik, I inherited the company's mission statement: 'Respecting the World, Respected Worldwide.' This ethos was founded with the aim of making our goods as energy efficient as possible. But as CEO, I wanted to widen this out, so that our sustainability goals would grow into every corner of the business. I knew that achieving this would require leadership on my part: if I could find a way to put myself out there, then others would follow.

At the same time, as the father of three young children, I was being asked questions about the environment I was finding increasingly difficult to answer. My eldest daughter, Andrea, now eight, is a big fan of skiing. But each year we went, she asked me about the changes on the slopes. Why is the weather warmer? Why is it raining on the mountain when it should be snowing? Sasha, two years younger, was similarly passionate about animals and animal welfare: not only was she not eating meat, she was quizzing me about why I was. My youngest, Oscar, was so instinctively in tune with nature, climbing trees to pick oranges and pomegranates, that I didn't want anything to sour that relationship.

Inspired both at work and at home about the planet, and infused with the personal challenge itself, in 2018 I decided to try and climb Mount Everest to raise awareness about the environment. The particular issue that awareness was focused on was the melting of the Himalayan glaciers. From Arçelik's business in the region, I'd become aware of the devastating effect that losing these glaciers would have: over two billion people relied on them for water, yet they had lost a quarter of their ice in the last forty years, at a rate that was increasing. I wanted to run a media campaign to raise awareness of this issue (and our videos would go on to garner 80 million views), but knew that we needed something eye-catching at the heart of the campaign to really grab people's attention. Climbing the highest mountain in the region – in the world – seemed the perfect way to go about doing this.

The only problem was that I'm not a climber. In fact, when I decided to commit, eight months before the expedition, I had never really climbed a mountain in my life. That experience of training and preparation, not to mention the climb itself, was the greatest physical and mental challenge of my life. I learned more about myself in the process than I ever expected: by pushing myself to the limit, I came back a different person and, I like to think, a better, humbler leader in the process.

This book, *A Mountain to Climb*, is partly the story of that extraordinary experience. But it's also the story of the planet, too, and the environmental crisis we find ourselves in. Over the last six years, I have tried to understand more about the issues we face and discover the possible solutions that could help us tackle them. I have spoken to environmental campaigners, young climate activists, biologists, scientists, film-makers, academics, economists, entrepreneurs, global leaders and innovators, among others, about what we can do. This book is an exploration of those discussions as well, the two narratives interweaving with the aim to both explain and, I hope, inspire.

'It didn't take long for the moon to become boring,' Bill Anders told the *Guardian* on the fiftieth anniversary of the Apollo 8 mission. 'It was like dirty beach sand. Then we suddenly saw this object called Earth. It was the only colour in the universe.' Speaking about his famous photograph, Anders said, 'It gained this iconic status. People realised that we lived on this fragile planet and that we needed to take care of it.'

Today, our planet is more fragile still. The need to take care of it is greater than ever. I hope in my own small way that the pages that follow will remind you why we have to do this, and suggest some of the possible ways that we can.

ON AN ISLAND

My journey to the top of the world started in a small boat on the Mediterranean.

It was summer 2018, and the living should have been easy. As had become our August routine, my family and I had decamped from the relentless buzz of Istanbul to the cool calm of the Aegean coast. When I was younger, I had a group of friends (still my closest friends today) who I'd regularly go sailing with. My friends would continuously press me to find land on the coast to build homes – the idea being once we had families and there were more of us, we could still spend the summers together. The plan was to anchor the homes with a resort that would at least partially cover the costs and take care of everything from the gardens to utilities and make it desirable to visit at all times of the year. For many years, I searched for a suitable piece of land but was always discouraged by the cost and bureaucracy of making something like that a reality. It was a pipe dream, after all. But then, completely by chance, I learned of someone my age who had done just that. Part visionary, part dreamer – with the required resources, of course – he had bought a huge piece of land on the Aegean coastline, against all conventional wisdom: he wanted to create a sustainable space where people could live closely in harmony with nature.

It was one of those rare moments in life that my dream intercepted with someone else's dream and my family and I became part of this wonderful community. Over the years, Kaplankaya became a haven, too, of festivals and ideas, a place for growth and development as well as relaxation and enjoyment. Today, it is both a community of old friends and a home for new ideas.

The ocean has always played an important part in my life, right from when I was a young boy. On this stretch of Aegean coast, the Mediterranean is at its best. It's the particular blueness of the water that always grabs me: its softness, its lightness, rippling and reflecting the peerless sky. That gentle movement beguiles. Sit on the shore and you can gaze for hours. But that's just the surface. Dive deeper, and it is a different story; there are darker tales of ecosystems fighting for survival and historical treasures of battles past. But above water, sat on the beach, all appeared calm.

That summer, I knew how that contrast felt. On the outside, my life was going well. In my career I had reached the position of CEO of Arçelik, one of the largest household appliance manufacturers in the world. I had the sort of job that was both stretching and stimulating. The market we were in was a highly competitive one, and I had both customers to please and a global workforce to look after. That was hard work and demanding: however many hours you gave to the job, it always wanted more.

Even when I was with my family, my beautiful wife Stephanie and my amazing children, I wasn't always *with* them. My head was somewhere else: full of work, preoccupied by planning and of organisation. Even at Kaplankaya, which was where I meant to go to relax, I still felt restless. With friends around, it's a social place, and I had a reputation for being in the thick of that. I was a life and soul guy: if there was a party happening, I was there. And if there wasn't a party, I was organising one. Rather than switching off and enjoying myself, all these social gatherings began to feel like just another set of dates to arrange, more events in the diary to go to. Life felt like a continual swirl of people, with no chance for me to catch my breath.

Inwardly, I felt sluggish. It was like wandering around in a haze, in a brain fog. I've never been a serious drinker, but I'm a social guy, and that socialising often involved a glass or two. Like many people, I'd been swept along by the resurgent trend for different flavours of gin. I liked a glass of wine with food, pinot noir and merlot particularly. And sometimes, after a meal, I'd order a Mezcal, a spirit made from the agave plant: it's smooth like tequila, smoky like whisky. I was self-aware enough to know exactly what that made me sound like, but that was where my life was stuck.

One evening, a good friend of mine called David pulled me aside. David is one of life's wise people. He's a thoughtful, spiritual sort of person, someone who has taken the time to work on himself. He's very grounded – one of those individuals with a low centre of gravity, if you know what I mean. When he pulled me aside, I was in my life and soul mode, as usual. It was another gathering of friends on the terrace as the sun went down, the buzz of conversation getting louder as the glasses continued to clink. David tugged me out of this, and led me to a quiet spot to speak.

'Listen, Hakan,' he said. 'I hope you don't mind me saying this, but I've been observing you for a while now. You're always in the thick of it. You bring people together non-stop. I really don't know how you do it.'

'Thank you,' I laughed. 'It's a difficult job, but someone's got to do it.'

'I didn't mean it as a compliment,' David said. 'All these people, all the time. Do you mind if I ask you a question? When was the last time you were alone?'

'Wow. Right. Well …' David's question pulled me up. I ran through my day to answer his question. In the car, I had a driver. In the office, there were hundreds of people waiting for an answer to this, a signature for that. Back home, I had Stephanie and the kids. Then there were the dinner parties, the social life, meeting up with friends to go diving or sailing or skiing or whatever. I tried to brush David's question aside. Of course, there must be times when I was alone. It was just that I couldn't think of any.

'I'm a Virgo,' I said. 'I like organising things.'

'And you're very good at it.' David gestured across to the drinks gathering that I'd arranged. Laughter and music echoed across the terrace. 'But, how can I put this? I worry sometimes that you're *too* good at organising things. That you might be doing all this organising not to please people, but as a distraction from yourself.' David shrugged. 'I don't know. Maybe I'm wrong on this. But as a friend, the fact that you can't tell me the last time you were alone worries me a little. And that should worry you too.'

David gave my arm a small squeeze and headed off. I stood there for a moment, watching him disappear into the fading light. Well that was a bit strange, I thought. Then I turned back to the terrace, and the noise of my friends. As I rejoined the group, a cheer went up, glasses were raised. But as I joined them, the ching didn't ring as clear as before. Now I could hear it more clearly it sounded dull, flat, empty.

• • •

'I don't understand,' Stephanie said, the following morning. 'But where are you going?'

'Boating,' I said, packing a bag. 'Just for a couple of days.'

'Who are you going with?'

'No one. Just me.'

Stephanie looked surprised at that. We'd been together long enough for her to know me inside and out. Like David, she knew how social I was by instinct, and so how unusual it was for me to do anything by myself. She knew, too, how much I instinctively relied on her. In so many ways, she was my rock and my support system. For me not to come to her, to step away, that meant that something serious was going on.

My children came rushing into the bedroom. They asked the same questions, and I gave them the same answers.

'But where are you going, Daddy?'

'I'm not completely sure,' I said. 'But I think I'll know when I get there.'

Children, as children do, took that at face value. 'Good luck!' They hugged me. 'We'll see you when you get back.'

As they ran out the room, Stephanie was holding up my wallet from the bedside table. 'Won't you be needing this?'

I shook my head. I reached into my pocket and pulled out my phone, putting it on the dresser. 'Or this.'

Stephanie looked at the phone. She knew how wedded I was to it. How messages would come in, disturbing our downtime, beeping and pinging in from morning to night. 'Are you OK?' she asked. I could hear in her voice how concerned she was.

I nodded. 'I will be,' I said, sounding more confident than I felt. All I could think of at that moment was to be alone, and to confront myself.

I went down to the harbour with the intention of borrowing a friend's boat. Hidden among the yachts and larger vessels, *Magnolia* was a small but beautifully crafted wooden boat, her mahogany sheen glistening in the morning sun. *Magnolia* was a modest offering compared to the other boats that were there, but there was something timeless about her design that I really liked. She had a real kick to her as well – her two engines could push up to thirty-five, forty knots.

As I made my preparations on *Magnolia*, the sun glinted off something in the water. I glanced down and saw a plastic bottle, one of those two-litre soda bottles, bobbing and floating its way inland. I shook my head: why do people throw away their rubbish like this? The Mediterranean is such a beautiful place – how can you treat it with such disrespect? I leant out of the boat to pull it out of the water. The bottle was half squashed flat, with its lid still on; its wrapper had deteriorated, so I couldn't work out what make it was. How long it had been in the water was anyone's guess. I chucked it down onto the deck: I'll dispose of that properly later, I thought.

I headed over to the kitchen, to prepare some food for the journey. Word had got out about my trip.

'Come on, Hakan,' one friend nudged. 'Where are you really going?'

'I reckon it's Mykonos,' another winked. 'Knowing Hakan, I bet he's got plans there waiting for him.'

I smiled and shook my head. 'Sorry guys, you're well off the mark with this one.' Apart from anything else, Mykonos was too far away on a craft that size. But even though the practicalities made their theory impossible, that didn't stop them speculating. I left them to it.

As I clambered back on board the boat and stowed my bag and supplies away, I paused over what my friends had said. Could I make it as far away as Mykonos? That would be a trip and a half. Or should I just drive around the corner, stop there and hang out? That would be too easy. No, I had a plan, and I was going to stick to it. I turned the engine on and, grabbing the boat's wheel, pulled out of the small harbour.

I had one final port of call before I could start my adventure proper. Didim. It's a place of important archaeological significance, home to the Temple of Apollo, one of the ancient wonders of the world; it's where Alexander the Great and the Persian kings came to seek advice from the oracle. My own visit there was rather more prosaic: to go on my journey, I needed to go through customs, and Didim was where I had to show my passport. The oracle told Alexander the Great to travel east. I, by contrast, was heading west.

• • •

My family has something of an international background. My grandmother is Greek, and my mother is Norwegian, but she was born in Turkey. By contrast, I was born in Norway but am Turkish.

I was born in Norway because that was where my family were based at the time. We lived there for the first five years of my life, coming back to Turkey when I was five.

In Norway, I learned to ski, but in Turkey I fell in love with the ocean. Each year of my childhood, we'd spend a month away on a small boat. Looking back now, it was an incredible thing to do. The boat itself was tiny: my father had built a sailboat out of a motor boat, putting in ballast and a mast. It was ten metres long and yet that became home to two parents, three siblings and a dog. Ten metres. How we squeezed in such a small space and survived for weeks, I'm really not sure.

We'd travel along the coast, go fishing and diving, explore the archaeological ruins. While my father's career had been in business, my mother had PhDs in archaeology and Byzantine art. She read Ancient Greek. Those sailing trips were full of stories from mythology, brought all the more alive by my mother's knowledge and telling, and by visiting the sites. As a boy, my head was full of these adventures, and I dreamt of the journeys I might embark on myself.

This childhood love of the sea blossomed in adulthood as my career took me around the world. For several years, work took me to Hong Kong. There was an embedded culture of sailing over there, yacht clubs with bulging memberships. That's how I spent my weekends: feeding off the competitive spirit on the water and enjoying the camaraderie in the bar afterwards. Living away from home, it was the sailing community that helped make Hong Kong feel a bit more like somewhere I belonged. From sailing locally, I then started travelling further afield. There were a whole raft of competitions and sailing regattas right across Asia I took part in. And then I started sailing around some of these remarkable places by myself: Burma, Thailand, the Vietnamese archipelago, the South China Sea.

I dived as well. I was fortunate that my chairman at the time was also into the sport, and he set up some remarkable trips that I was able to tag along on. He'd hire marine biologists and *National Geographic* photographers and get us out to places far off the beaten track. We'd moor a boat and dive down to these remarkable underwater worlds. There are so many places I could mention, but visiting Layang Layang, an atoll of thirteen linked coral reefs off the Borneo coast, is one trip I'll never forget. The water was crystal clear – the visibility can reach forty metres – and the richness of the fish species and coral diversity are remarkable. There's something, too, about experiencing a school of hammerhead sharks swim past that lingers long in the memory.

Layang Layang is one of those rare areas that has never been fished in, and remains relatively free of pollution. It's a reminder of how the oceans, all the oceans, used to be. As well as infusing me with a love of the sea from an early age, my parents, and my mother in particular, made clear the environmental impact that man had on the sea. As well as her archaeological knowledge, my mother was also one of the founding members of Turmepa, an NGO set up to help preserve the Mediterranean, and particularly the waters around Turkey. Even when I was a small child I'd go on beach clean-ups, clearing up the trash. The sea itself wasn't as polluted as it is now. We'd drop a net every night, catch fish to eat. There wasn't the plastic and the rubbish you'd pick up now. But there was still plenty being washed up on the beach, even then.

My mother fought hard. She believed that the best way to counteract rising levels of pollution was to educate children, entrenching environmental beliefs from an early age. She created books about a group of sea creature characters, and the effects of throwing waste into the sea. These were distributed to millions of children right across Turkey. The group took on the polluters as well, tankers who were discharging their dirty water into the sea, and factories that weren't filtering their discharge; the group even employed boats that would travel from yacht to yacht, collecting people's rubbish so they didn't throw it overboard.

It was an environmental message that I took on board from an early age, and a belief that has stayed with me ever since.

• • •

From the shore, the Mediterranean appears smooth, like a millpond. There's something about the rich shade of blue, the sky reflecting off, that beguiles and tricks the eye. But once you're out into the open sea, especially on a boat as small as *Magnolia*, it can be a completely different story. I knew from experience that the swell in this part of the Mediterranean can engulf a small craft, with four-metre-high waves that can sweep a boat away. Away from land, you're aware of how the sea is this endless, living, breathing thing, how large and powerful nature is, and how small and insignificant you are in return.

The more the Turkish coastline faded away into a thin strip of silhouette behind me, the stronger the wind got, and the more the boat smacked down onto the waves. The plastic bottle I'd picked up bounced up and down on the deck and I could feel my hand gripping the ship's wheel tighter, to make sure I remained in control. The sea spray became relentless. I had to put on my diving goggles so I could see where I was going. But I didn't stop. I'd come too far to turn back. There was something about the moment, the sudden emptiness, that drove me on.

Ahead of me, in the distance, I could see the outline of Arki. Arki is a small Greek island, one of many in the Dodecanese archipelago. It's secluded and unspoilt, has 150 or so inhabitants, but even that was too many for me. Instead, I drove the boat towards one of the outcrops nestled near the island: a place I knew well and loved.

Four years previously, I'd come to the outcrop with Stephanie and a friend of mine, Burak, to go paddleboarding. Stephanie was pregnant at the time with our son Oscar, though that didn't slow her boarding down – in fact, she left me and Burak for dust. The outcrop is one of those magical places to be. The water is so clear there you can see down and down into the blue. If you know where to look, it's got a cave-like entrance, where if you dive down, you can find a tunnel underwater. Swim through that, and it opens out into an extraordinary lagoon. It's perfectly formed, cocooned by steep sides. The sides are about ten metres tall; you can climb them and dive in. I don't know enough about geology, but it looks like it was hit by a meteor millions of years ago. It's one of the most beautiful places I know on earth.

On the way back to the boat on that trip, Stephanie had raced on ahead. And as Burak and I followed on our paddleboards, I noticed something move out of the corner of my eye. I glanced across, and less than five metres away, saw a large grey shape on top of a rock. It was so big that my initial thought was that it was a dead horse, somehow washed up and beached there. But then I saw the whiskers, and these huge eyes swivelled round to look at me.

'Burak,' I hissed. 'Look at this!'

The creature was a Mediterranean monk seal. They're incredibly rare – there's less than a thousand left in the world, and only a few hundred in the Med itself. To see one, and so close up, was an extraordinary privilege. Rather than swimming off at the sight of us, the monk seal stayed. It slid off the rock and swam around with us. It was intrigued by us, interested and playful. By the time we got back to the boat, Stephanie was wondering what had happened to us. She didn't believe us at first about what we'd seen. But we have called the island Seal Rock ever since.

Four years on, I arrived at Seal Rock by myself. No wife, no friends, no seal. As I cut the engine, I became aware of the quiet. When I dived down, found the tunnel and kicked through to the lagoon, it felt quieter still. I was properly alone, with no way of contacting anybody. I kicked around the lagoon for a bit, but it felt different without anyone else there. I thought I might find some peace in the silence, but instead, it just felt empty. I headed back to my boat and went on my way.

I drifted around, until I found another island, one with a bay big enough for me to anchor. Seal Rock was stunning, but it wasn't somewhere I could leave the boat, especially not with the way the wind could pick up. The bay I found on the next island, however, was far more secluded. It was sheltered and shallow: the perfect place for me to stay overnight.

The only thing stopping it being perfect was that I wasn't the only person there. On the other side of the bay was another solo sailor, one of those nomadic sorts who travel from island to island, traversing the world. He was very much of a type – heavily tanned from all the fresh air and sunshine, tending a ramshackle boat powered by sails and solar panels. We acknowledged each other, but that was as far as it went – he wanted to be left alone as much as I did, which suited me just fine. The fact that he was there and seemed to be at peace with himself encouraged me. It felt as though I'd ended up in a good place.

Having arrived, I wasn't quite sure what to do with myself. Bereft of books, of phones, of music, I had nothing to distract myself with. I didn't need to cook, as I had all my food pre-prepared. I realised, and this seems ridiculous, that I didn't even have a pair of shoes with me, so couldn't go and properly explore the island. All I had was the sea and my mahogany boat. I went for a dive, found some sea urchins to eat for later. But I didn't feel hungry. Back on the boat, with nothing else to do, I started to clean it.

I was incapable of sitting still. As I continued to clean and busy myself, rather than making me feel better, I realised I was feeling worse. It felt like a swell inside, like the waves of the ocean I'd smacked into on the way here. But this wave, rather than rising and falling, felt as though it was getting larger and larger. And then, in a rush, that emotional dam burst. I could feel my eyes pricking with tears, and though I tried to blink them off, they kept on coming. And then I found myself crying, really crying. It can't have been a pretty sight, and I'm glad there was no one there but a disinterested sea nomad to witness it. Because once I started, I really let it all out. There was no filter or self-consciousness. I cried like I hadn't cried in years. In fact, I couldn't remember the last time, if ever, that I'd cried like that.

The sea nomad hadn't been disturbed by my outburst, but something else had. As I looked up to the heavens, I saw a shape in the sky. Through teary eyes, the silhouette looked blurred, but as I rubbed them, I realised what I was looking at. An eagle. It was swooping from side to side, its wingspan stretched wide, its dark brown feathers the colour of my boat. It was observing me, looking down just as I was staring up at it. With each sweep of the sky it dropped lower and closer – so close that I could see the swish of its feathers in the wind. I don't know how long we watched each other, but it felt a long time. By the time it gave a powerful pull of its wings and weaved itself up and away, I found myself noticing the silence. I'd stopped crying. The swell from within had gone. I felt as though something had shifted.

I stayed in that bay for the best part of two days. I did a lot of thinking. About myself, about my life. I need to stop drinking,

I decided. I wasn't a drinker drinker, but I was drinking enough for it to make a difference. That needed to stop. The endless socialising, too, I realised wasn't so much about seeing other people as avoiding being by myself. It wasn't a bad life – don't get me wrong, a lot of it was fun. But it was empty fun. Hedonistic. Selfish, if I was being honest.

I didn't like myself, the person I'd become. I realised that with a lurch. There wasn't enough meaning in the way I was living my life. For all the work I was doing and all the career success I'd had, I wasn't building enough value. It was as though I was coasting through life, not properly stretching or testing myself. Was this what a midlife crisis was? That certainly crossed my mind. I'd watched friends go through that, usually with the same self-destructive results. Divorce, drinking, reckless spending. I didn't want to do any of that. But I realised I needed to do something.

I tried to take a break from myself, to take in my surroundings. There was a stream running off the rocks into the water. I followed that down into the bay, the rich carpet of seagrass bending and waving underneath the surface. The sea life was rich in abundance. I saw a turtle. Everything felt in harmony, the opposite to how I was feeling.

But my mind was beginning to whirr with thoughts. The turtle, the eagle, the monk seal on Seal Rock. Memories flashed of the coral reefs I'd scuba-dived when I was living in Asia. The childhood trips with my parents along the Aegean coast. The constant, I realised, was nature. It was something that gave me joy, yet like so many people, it was something I took for granted. I glanced down at the plastic bottle I'd picked up back on shore. It wasn't just me who was out of kilter with the world. We all were. It wasn't just me who needed to change. We all did.

But what could I do to make a difference, a real difference? I tried in my day job to instill change, but that was hard work. It felt at times a constant battle, a lonely path. But what if I could find a way to inspire my colleagues to show how far I was prepared to go? I had an inkling of an idea. Two friends of mine were talking about trying to climb Everest. I'd laughed at them when they mentioned it, but they were serious. *You should join us, Hakan,* they'd said. I'd laughed at that too.

I wasn't a climber. I'd never climbed a mountain in my life. And yet in that moment, part of me was drawn to the sheer audacity of the idea. Maybe a longshot expedition was exactly the sort of thing that I needed to shake myself out of my sluggishness. Maybe, too, I could use such an expedition to help raise awareness about the environment. It was one thing talking to employees or fellow business people about how to do better for the planet. But if I could put my money where my mouth was, and pull off something like this, then they'd know how serious I was. Maybe then, people would sit up and take notice.

Sitting on the deck of *Magnolia*, this small boat floating in the Mediterranean sunshine, the thought of climbing the highest mountain in the world seemed far-fetched and preposterous.

I couldn't.

Could I?

FEELING THE HEAT

On the opposite side of the Mediterranean, the small coastal town of Mati is found on the Attica peninsula of mainland Greece. Thirty kilometres from Athens, it is both a popular tourist resort and one with plenty of second homes owned by those escaping the capital for the weekend. Close to Rafina, a harbour town which serves as a port to the Cyclades islands in the Aegean Sea, it is full of everything you'd expect from a Greek seaside location: apartment blocks and villas, tavernas, bars and cafes, sandy beaches and marinas.

But in July 2018, a month before my sailing expedition, Mati became internationally known for a different, tragic reason. As wildfires from that summer's heatwave spread across Greece, it was Mati that suffered the most.

The fire was brutal. At 6 p.m. on 23 July, smoke reached the town. By 7 p.m. the flames were engulfing properties, with the sky turning a vivid red. By 8 p.m., the fire had destroyed the town. 'Mati no longer exists, said Evangelos Bournous, the mayor of Rafina. Powered forward by gale-force winds, it is estimated that the heat reached temperatures of up to 800°C. The fires burnt down 40,000 pine and olive trees, together with 4,000 homes and hundreds of vehicles. Rescuers compared the scenes of devastation to a 'Biblical disaster' or a conflict zone similar to those in war-torn Syria.

But it is the human cost of the events of that summer evening that's most tragic: the disaster left 103 people dead and hundreds of others fighting for their lives in hospital. The stories of those who survived the ordeal are terrifying. Dimitris Matrakides ran into the sea to try and escape. He was one of hundreds who spent seven hours in the water, waiting to be rescued. As molten ash rained down, debris flying through the air, and gas bottles from beach cafes exploding, he stood in the sea up to his neck, hoping. 'I still have nightmares,' he told the *Guardian* newspaper a year later. 'I am always in the sea – my legs freezing in the cold, my torso frying in the heat.'

Another resident, Kalli Anagnostou, remembers the thermal wave that hit before the flames appeared. 'It was so violent, with explosions, corpses, cars, houses and trees on fire – and everywhere black, toxic smoke.' Her son's flip-flops melted in the heat as they escaped the flames. Kalli suffered fourth- and fifth-degree burns across 40 per cent of her body, and was in a coma for three weeks while the doctors carried out plastic surgery. She will continue to need operations for years to come.

But for all their horrifying experiences and injuries, the survivors of the fire are the lucky ones. Another resident, Barbara

Kasselouri, told the *Guardian* how a neighbour, together with her daughter and granddaughters, all died: 'Her last words on the phone to a friend were: "The fire is coming, we don't know what to do."' With the fires blocking roads and paths out of the town, heading to the sea was the only option. One group found their way blocked by a burning tree and doubled back to find another route. They ended up on a clifftop with no apparent way down among the steep rocks to the beach below. Rescuers later found the bodies of twenty-six people here. Overwhelmed by the flames, the bodies were found huddled together, an instinctive human reaction to face their final moments in support of each other.

. . .

In the aftermath of the fires, investigators laid part of the blame for the disaster on human behaviour. It was thought that a sixty-five-year-old man had unwittingly started the fire by burning wood in his back garden. But the lack of communication and poor response of the emergency services also played their part as, further back, did planning decisions that allowed properties to be built so closely together. In March 2019, Greek prosecutors charged twenty people on counts of arson, manslaughter and grievous bodily harm due to negligence.

Human behaviour, however, was only partly responsible for the fires. Over the previous year, unusual weather conditions had created the tinderbox situation the town found itself in that July. The previous summer had also been exceptionally hot, and was followed by an extremely dry autumn. With groundwater sources unable to be sufficiently replenished, the area entered 2018 having still failed to recover from 2017. So when another heatwave hit that summer, the conditions for fires were already in place.

It wasn't just Greece that was hit by another heatwave in those summer months: across the northern hemisphere, records were tumbling. The same week those individuals lost their lives in fires in Greece, Sweden was appealing for international help in dealing with wildfires there, including a dozen raging in the Arctic Circle. In Ouargla, Algeria, a temperature of 51.3°C was recorded, the highest ever measured in Africa. In California, increasing use of air conditioning to deal with the heat resulted in power shortages. In Montreal, Canada, the city's morgue was overwhelmed with the bodies of people dying from the heat.

Meteorologists put the hot temperatures partly down to the jet stream – air currents which help to push the weather around the world – being weaker that year. This meant that once hot high-pressure air was in place, it stayed there, rather than moving. Major changes in the sea surface temperatures of the North Atlantic – the Atlantic Multidecadal Oscillation, or AMO – also played their part.

But while changes to sea surface temperatures and jet stream strength have changed in cycles over the centuries, the combination in 2018 was exacerbated by the role of carbon emissions in raising baseline global temperatures. The year 2018 was, at the time, the fourth hottest year on the planet since records began. The previous three years had been the three hottest. Since 2018, both 2019 and 2020 have been hotter. In January 2021, it was revealed that 2020 had tied with 2016 to be the hottest year on record. And that was despite both the global economic shutdown due to the coronavirus pandemic and a La Niña weather event, which naturally lowers temperatures.

In the final days of July 2021, wildfires broke out in the coastal towns in Turkey – Marmaris, Bodrum, Manavgat and others. Over 100 wildfires were recorded in a span of days. As intervention was delayed and human fallacy appeared, the fires raged on and on, reaching heights that were no longer manageable for the local forces.

It is no surprise that Turkey witnessed the worst wildfires in its history that summer – nearly 178,000 hectares of forest had already burned by August; eight times the average between the years of 2008 and 2020.

As Turkey struggled to breathe over the flames, more than 800 wildfires were recorded in Italy as resort towns were being evacuated. Greece and Spain witnessed temperatures above 40°C, with fires picking up, forcing them to evacuate thousands of citizens and tourists alike.

According to US officials, in the final days of July 2021, 90 fires were burning in twelve states, and have already destroyed 1.8 million acres. In October 2020, the US recorded its first 'gigafire' that burned over one million acres in California. One of the most notoriously cold places in the world, Siberia, recently saw wildfires so intense that the smoke reached the North Pole. Russian officials state that the fires of 2021 ravaged more than 14 million hectares, making it the second-worst fire season since the turn of the century.

Out of the ten hottest years on record, nine have happened since 2010 (the tenth occurring in 2005). The summer of 2018, and the wildfires of Greece, as tragic as the consequences were, were in fact relatively unexceptional. Over the last decade, extreme weather events have become the norm. I could have begun this chapter by talking about the East African drought of 2017, which saw 13.1 million facing famine levels of food insecurity. Or the water shortages in Bolivia in 2016, the worst the country had faced for twenty-five years. Or the Pan-Caribbean drought of 2013–16, the worst in the region since 1950. Less known but every bit as important was the marine heatwave of 2016, which affected temperatures on the ocean surface, and led to the mass bleaching of coral reefs in Australia.

In other parts of the world, flooding and extreme rainfall has been the issue. This was the case in south-east China in 2015, Bangladesh in 2017 and Nigeria in 2018, to give just three

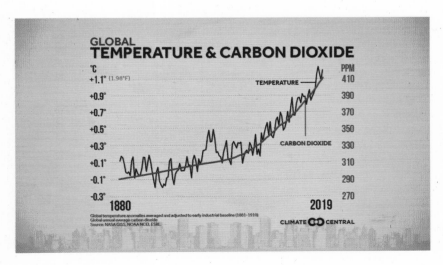

Global temperatures have risen around 1°C since the 1800s. (Source: NASA)

examples. Elsewhere, that precipitation led to heavy snowfall: in Nepal in 2014, a blizzard dropped over 1.7 metres of snow in twelve hours, leading to avalanches and subsequent deaths. In 2017, the US recorded sixteen billion-dollar weather events at a cost of $306.2 billion: Hurricane Harvey, which cost $125 billion, was the single most significant tropical cyclone rainfall event ever recorded in US history. In 2021, we witnessed how vulnerable developed economies were against extreme weather events: even on the European continent, where natural disasters are few and far between, extreme flooding took over 200 lives.

· · ·

I grew up on one of the Princes' Islands, an archipelago of small islands in the Sea of Marmara, a few miles off the coast of Istanbul. It was an idyllic place to be brought up and special in many ways. There are no motorised vehicles allowed on the islands, just horse-drawn carriages and bicycles to get yourself around. The architecture is stunning – you're surrounded by buildings that are two or three hundred years old. For a child growing up, it's an amazing place to call home: quiet, unspoilt and safe.

The population on the islands has generally been made up of minorities in Turkish society: Greeks, Armenians and Jews. My family had a long heritage of using those islands as a summer residence. Those working would get the ferry across to the mainland, and the rest of the family would stay on the islands.

That's where I spent my summers growing up. I used to go fishing with local fishermen very early in the morning: we'd fish from four in the morning until ten, and then sell the catch at the market. That taught me how to fish, but was also my first experience in buying and selling – the fishermen would use me as a small child to sell their wares! Even at that early age, I was quite good at striking a deal.

It was an idyll that didn't last forever. Spending that much time on the sea, it soon became obvious to me how things were changing: the pollution in the water, the plastic and other detritus, and how the population of fish was starting to shrink. There was a time when we could drop a net in the water right in front of the house and get an abundance of goods that we would share with friends and family. To begin with, the net would be bulging with so many different species of fish. But as the years went on the net would get less heavy, and the catch would be lighter.

Istanbul changed as a city as well. It grew from a city of under 3 million of my youth, when I was small to the giant it is today, housing 15.5 million people and still growing. Where once you could look across to rich green hills and empty pastures, construction has turned it into concrete as far as the eye can see. That rise in population has been equalled by a rise in traffic and a subsequent deterioration in air quality. During the pandemic lockdown in spring 2020, it was noticeable how the air quality improved as traffic disappeared, and how it dropped again when restrictions were lifted and traffic returned to normal.

The weather has changed as well. Back when I was a child, Istanbul would play host to four distinct seasons. Winter meant snow. Most years we'd have snowfall for at least a week, a couple if we were lucky: the schools would be closed and we'd go out and play. We'd head out of Istanbul, two hours up into the mountains, and go skiing. People often don't think of Turkey as a snow country, but 70 per cent of the land is on an elevated plateau and has a snow footprint (that's why, historically, Turkey has been a great place for farming and food production: the snow seeps into groundwater, which makes the land fertile). Today, all of that has changed: snow in Istanbul is a rare occurrence, and the four seasons blur into one.

That's my anecdotal experience as someone who has watched a city grow and change over their lifetime. But the evidence is there to back this up: one research paper published in 2017 shows that between 1912 and 2016, the average annual temperature in Istanbul rose by 0.94 degrees. This rise correlated with the industrialisation of the 1940s and later with the growth in population and decrease in vegetation cover as the city expanded. At the same time, precipitation in the city also increased, as did the frequency of heavy rainfall.

Wherever you call home, I'm sure you can think of changes in the climate in recent years. All of us who are alive now are witness to how the world has altered, and can see the evidence of climate change with our own eyes. It's a process that has accelerated over recent years, and the last decade in particular. But it's one that began much further back – the consequence of human behaviour that stretches over recent centuries.

• • •

Svante Arrhenius was something of a maverick scientist. Born in Sweden in 1859, he was prodigiously talented, particularly in arithmetic. But studying science at university, his doctoral dissertation on electrolytic conductivity confused those marking him, and he was awarded the lowest possible grade. It turned out he was just ahead of his time – twenty years later, he built on the same ideas to win the Nobel Prize for Chemistry.

But it was another piece of scientific theory that Arrhenius completed in 1896 that, although he received no Nobel Prize for it, was perhaps the most significant work of his career. Arrhenius was intrigued by a contemporary scientific debate about what the cause behind the ice ages was. Earlier scientists, such as Joseph Fourier, had already discovered the role of carbon dioxide, and how atmospheric gases affected the temperature of the planet. CO_2, researchers were beginning to realise, trapped the sun's energy in the atmosphere, rather than releasing it and allowing the atmosphere to cool. At the same time the resulting warmer air held more water vapour, which in turn magnified the heating effect still further.

The question Arrhenius asked himself was whether he could calculate the levels of CO_2 and water in the atmosphere that heated the planet – and in terms of the Ice Age question, how far levels would have to fall to lower the temperature (in the late 1800s, this was more of a live issue, with the world having lived through the Little Ice Age from the sixteenth to the mid-nineteenth century). Arrhenius spent a year doing what he described as 'tedious calculations'. He did a lot of these – tens of thousands – before he came up with his conclusions. If CO_2 levels fell by half, he deduced, then that would lead to a temperature drop of 4–5

degrees. If CO_2 levels doubled, conversely, then global temperatures would increase by 5–6 degrees.

Arrhenius's calculations were the first time anyone had
attempted to work out what we now call the greenhouse effect.
Considering that his sums were worked out with a pen and paper,
rather than using computer modelling, he wasn't far off. Today's
estimates are that a doubling of CO_2 would lead to a 2–3 degree
rise, rather than the 5–6 degrees he suggested.

In his calculations, Arrhenius got one other key fact right,
and one wrong. By the end of the nineteenth century, the world
was in the midst of what was considered the Second Industrial
Revolution. The use of steam engines and the burning of coal
had rocketed over the previous hundred years: in 1800, the world
production of coal stood at 10 million metric tonnes. A century
later, it was 1,000 million. Arrhenius understood that the burning of fossil fuels would increase the amount of carbon dioxide
that ended up in the atmosphere. But where he got it wrong was
how quickly coal burning would cause these temperatures to rise.
His estimate was that coal burning would cause a 50 per cent
rise in CO_2, but that this would happen gradually, taking about
3,000 years to do so. In fact, by 2015, the amount of CO_2 in the
atmosphere passed 400 ppm (parts per million) – already a 40 per
cent rise from pre-industrial levels (280 ppm). In less than two
centuries, we are already 80 per cent of the way to what Arrhenius
predicted would take three millennia. To give you some sense of
scale, the last time there was as much CO_2 in the atmosphere as
there is now was several million years ago.

What Arrhenius (not unreasonably) failed to predict was how the twentieth century would unfold: the level of economic growth, the rise in the global population and the increase of energy use. The numbers, particularly in the second half of the century, are simply extraordinary. In 1820, there were approximately 1 billion people on the planet. In 1900, there were 1.6 billion. By 2000, that number had grown to 6 billion. James McNeill puts it even more strikingly in his book *Something New Under the Sun: An Environmental History of the Twentieth-Century World*: 'Another way to conceive of the extraordinary demographic character of the modern era is to estimate how many people have ever lived ... about 80 billion hominids have been born in the past 4 million years. All together, those 80 billion have lived about 2.16 trillion years ... Although the twentieth century accounts for only 0.00025 of human history, it has hosted about a fifth of all human-years.'

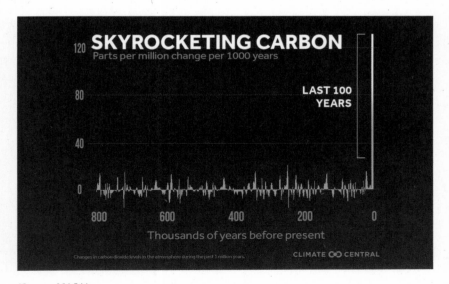

(Source: NASA)

The numbers on economic growth and energy use are no less striking. According to Our World in Data, using numbers adjusted for inflation and expressed in 2011 prices, the size of world GDP in 1500 was about $430 billion. By 1820, this had tripled to $1.2 trillion. By 1900, the effects of the Industrial Revolution had seen it almost triple again, to $3.42 trillion. In the first half of the twentieth century, this rise continued: by 1940, world GDP stood at $7.81 trillion. But after the Second World War, the graph of growth is almost a straight line. By 2015, the figure stood at $108.12 trillion.

The energy figures are no different. The production of coal, as mentioned earlier, went from 10 to 1,000 million metric tonnes over the nineteenth century. By 2000, it was five times larger again. Oil production, non-existent in 1800 and only 20 million metric tonnes in 1900, stood at 3,000 million in 2000. According to one estimate, we used more energy in the twentieth century than in the entirety of human history before it.

• • •

Given all that activity – the staggering growth of the world population, the resulting levels of consumption, the sheer amount of products made, and the energy used to facilitate all of this – the increased levels of CO_2 in the atmosphere feels almost inevitable. For a while, scientists wondered if the oceans might be able to soak up the additional CO_2 that the growing population were producing. It wasn't until the 1960s that researchers concluded this wasn't possible. Even then, the predicted growth in temperatures was over a long time period, if shorter than Svante

Arrhenius's original guess: new models suggested a 2°C rise over hundreds of years. As the century continued and the evidence mounted up, so the estimated time period for such temperature rises continued to shrink.

Since pre-industrial times, the planet has already warmed by over 1°C. On current projections, it is likely to rise above 1.5°C between 2030 and 2052 (according to the latest World Meteorological Organisation's five-year prediction, there is 40 per cent chance global temperatures will exceed 1.5°C in at least one of the next five years). Some additional rises in temperature are already baked in: there is a time lag between greenhouse gas emissions and their effect on the climate. So the amount of leeway we have to avoid a 1.5°C increase is already incredibly small.

Why is 1.5°C such a crucial figure? If you think the weather is getting more extreme now, it's nothing compared to what happens above this level. Once you're over 1.5°C, the effects of climate change begin to seriously ramp up. As soon as the 1.5°C temperature threshold has been passed, change is no longer steady – everything starts to accelerate. Back in 2018 the Intergovernmental Panel on Climate Change (IPCC) published a report called 'Global Warming of 1.5°C'. It's not a report to read before you go to bed, outlining in stark terms the different effect on the planet between a 1.5 and 2 degree rise in temperatures: ever-increasing heatwaves, droughts and flooding; sea levels rising, tens of millions hit by coastal flooding; a staggering loss of biodiversity; coral reefs completely wiped out; 2.5 million square kilometres of permafrost thawing which, in turn, releases methane, another even more potent greenhouse gas. The same panel then released another report in 2021, stating that humans are 'unequivocally' responsible for the climate crisis.

For politicians, for policymakers, for business leaders, for all of us, the challenge is finding a way to play our part in keeping those temperature rises below 1.5°C. To do that, current CO_2 emissions will need to have declined by 45 per cent by 2030 and reach net zero by 2050. Even with the ambitious pledges of COP26, our short-term goals still put us on track for a 2.4°C increase. I think back to those twenty-six people trapped on that Greek clifftop in 2018 as the wildfires swept in. If we don't act, then that is a metaphor for what may follow: humans as a species huddling together, unable to survive as the planet heats up.

What makes that particular story all the more heartbreaking is that there was a way down from the cliff. The land they'd ended up on had been in the same family's hands for decades. Years before, the present owner's grandfather had cut a rough set of steps into the cliff face, which took you down to the rocks below. If you knew where the steps were, there was a way to escape. But if you didn't know they existed, you'd assume you'd reached a dead end.

That's where we're at with climate change right now. Yes, temperatures are rising to the point of getting out of control. But there is still, just, a chance of escaping the deteriorating situation we find ourselves in. That pathway is there: it's not an easy one to find, or an easy one to take, but it does exist. Over the course of this book, I want to introduce you to some of the amazing guides, and some of the innovative ideas, that could help us get there.

But before we get to that, I need to tell you about my own journey, and my attempts, in my own small way, to do what I can to make a difference. It's a story that started just a few weeks after those terrible events in Greece, on another shoreline further along the Mediterranean.

MISSION ACCEPTED

The deep thud of the Lama helicopter's rotor blades was so relentless that it hid the thumping of my own heart almost entirely. I was high, high up above India's Himachal Pradesh, with the rounded glass panels of the helicopter cabin giving me a goldfish bowl's view of the landscape below. This northern part of India is an area of extremes: plunging river valleys and vertiginous peaks. Snowfall interweaved its way up the black rock of the mountainsides, their monochrome colours contrasting sharply with lush greens and browns of the lower hills. Beyond, the blue sky tinged with pink, the knife-sharp ridges of the Himalayas scored the horizon.

This was the craziest part of a crazy week. I'd arrived in the noise and heat of Delhi a few days before. From there, I'd taken the train up towards Manali, a two-day trip that allows you to observe one of the world's great countries through the carriage window. Inside the train, the cabins were crammed with people, chickens and goats, with many more passengers on top of the train, like a scene from a movie. Among those making the journey were many young couples: Manali is one of the honeymoon centres of India. It's a well-worn journey: get married in the south, escape to the north. The train ride ends short of Manali; there's a further eight hours' drive to get there. But when you finally get there, you can immediately see why it is popular: the scenery, the freshness of the mountain air.

Back in the 1960s, Manali became a bit of a pilgrimage for the counterculture. It had a reputation for offering some of the world's best hash. Even today, you can head down to the town's bazaar and see the blue-eyed sixty-somethings in full Indian outfits, the ones who had such a great time they never left. More recently, the region has used its natural setting to pitch itself as a sort of adult adventure playground. Trekking, climbing, white-water rafting and paragliding are all popular. But above all, the region has become an increasingly enticing place to come skiing. On the train, among the honeymooners, I noticed fellow travellers lugging the same telltale long, thin, black bags. There was a nod of acknowledgement between us. Here was a kindred spirit, someone else who knew why they were heading for Himachal Pradesh and what exhilaration awaited.

Skiing and winter sports have always been in my blood. I spent the first five years of my life in Norway, as my parents were studying for PhDs there. As a baby, they would put me into a sheepskin sled and take me langlauf, or cross-country skiing. Some of my earliest memories, before we moved back to Turkey, are all about the white and the snow.

Over the years, my passion for extreme sports has taken me skiing in some amazing places. In Georgia, I discovered the beauty of the mountains on the Black Sea coast. Turkey, if you know where to go, has much better skiing than you might at first think. The way the weather system works means that there's a lot of fresh water evaporating, which in turn leads to regular falls of fresh snow. It's light snow, which means that it is safe and predictable. On one trip, I even went skiing in Greenland. It was incredibly cold and icy, but with amazing views of the sea – it's quite an unusual experience.

In Canada I first discovered the joys of heli-skiing. Heli-skiing is where you are flown to remote mountains and slopes that you wouldn't normally be able to access. This is a real test of your skiing ability – every peak is different, and the snow is pure and untouched. Rather than following the established routes of previous skiers, you're following guides down a different route for each run. It's exhilarating. There's always the risk of avalanches, so you have to be careful. But the feeling of freedom and of being in nature is unparalleled.

Then I heard about a guy called Roddy Mackenzie. Roddy is a New Zealander guide who runs heli-skiing trips in India. He is an inventor, too: he came up with an adapted Jeep that you can use in the mountains and at high altitude. He also created a pressurised sleeping bag for people who got altitude sickness, which he sold to the Indian army. And he'd set up a heli-skiing operation that made an innovative use of RF radio tags – these were handed out to his skiers, so that if anyone was in an avalanche, the helicopter would be able to return and find them.

My helicopter thudded on. Four of us sat in the back: three skiers and one guide. The guide, like all of Roddy's, was a veteran skier. All selected from New Zealand, Austria or Switzerland, they were among the most experienced in the world. But even this guide didn't have any experience of where we were – this was the first attempt at a new descent, from 5,600 metres up. Back down in Manali, the four of us had looked at the proposed route on a laid-out map with a mixture of whistles through teeth and intakes of breath. Someone had suggested, I think it might have been my friend Nicholas Reille, that we nickname the route the PP, because we were all peeing our pants about doing it. Everyone had laughed at the time, and the nickname had stuck. Now, as the helicopter flew closer to our drop-off point, the reality of what we'd agreed to was beginning to bite.

With a grin, the guide reached into his pocket and pulled out four matches. One of these he snapped in half. Then, with a bit of careful concealing, he arranged the four matches in his hand, so just the tops of each were visible between thumb and fingers.

'To see who goes first,' he shouted above the noise of the helicopter blades.

He offered his hand to the skier sat on his left. The skier slid off his glove, staring intently at the matches. Then, reaching forward, he pulled out one of the two middle matches.

Long.

The guide leant forward, held his hand out to Nicholas, who was sat next to me. He wavered for a moment, then pulled out the remaining match in the middle.

Long.

The guide smiled across at me.

'Looks like it is one of us, then.'

I grimaced. Outside, I could see the helicopter was climbing to where we'd be setting off from. The mountainsides, so awe-inspiring as a backdrop, were beginning to feel a bit too real. Close up, they looked steeper too. Just to add to the tension, the higher the helicopter got, the more it started to shake, as it struggled with the altitude. At times it felt as though the helicopter could barely hang in the air.

I looked down at the two remaining matches in the guide's hand. Here goes nothing, I thought. I picked the one on the left. I teased it out, hoping to see the long slow draw of the full match. But barely had I pulled on the match before I could see its splintered broken end.

Short.

Shit.

The other two skiers whistled. The guide opened his hand up to reveal the remaining long match. He shook his head and tried to hide his grin. He was about to say something when a knock from the pilot told us to get ready to depart.

'OK, Hakan,' the guide said. 'Time for the PP.'

I could feel the adrenaline pulsing through me. I'd done a lot of skiing, but even with all that experience, I could feel I was shaking a little. The helicopter, too, was really struggling now, with the crystal blue of sky disappearing as the white of the mountains filled the windows. Beneath, I could see the surface layer of snow whirling up in the downdraft of the helicopter as we locked in. I checked my gear, that I had my avalanche transceiver and it was working. I tightened my goggles, then the straps on my gloves.

The helicopter was coming in close now. I could see the small ledge that the pilot was aiming to drop us off at. I was grateful for the fact that the wind had dropped, and the day was remarkably still. I was less grateful for the shute I was looking down at. The ledge was, at most, a metre and a half wide, and the angle down from it was about 45 degrees. Looking further down, I could see a sheer drop of about 1,000 metres from the edge of a cliff below. If you didn't turn hard right in time, then you'd be straight over the edge.

'Time,' said the guide, sliding open the side door of the Lama. Even without any serious wind, the power of the engine was more than enough for a blast of ice-cold air to hit me in the face. Flakes and flecks of snow whirled all around – it was like stepping out into a snow globe.

One by one, we jumped down from the helicopter onto the ledge. The guide went first, grabbing the skis and poles from the basket on the side of the Lama. Then we followed. The snow was deep enough on the ledge that my boots sank in, cushioning my fall. With a thumbs-up, the pilot pulled the helicopter away. The constant thrum of the blades faded and then disappeared. As the helicopter vanished, descending sharply down and out of sight, all was silent. There was no going back now.

The four of us clipped into our skis. I checked my goggles, gripped tight on my poles.

'Good luck!' The guide gave me a friendly pat on the arm. 'Remember. Hard right.'

I gulped, looked down. Here goes nothing, I thought. I launched myself forward, and felt the cold on my face as that familiar rush and sense of speed started to kick in.

• • •

'The only point I've ever seen for climbing a mountain is so you can ski back down,' I said. 'I've never understood why people would climb a mountain to then climb back down again. It doesn't make any sense.'

On the opposite end of the phone line, Lukas Furtenbach sounded nonplussed.

'Right,' he said, after a pause. 'Do you have any previous mountaineering experience?'

I shook my head. I listened to Lukas as he made a note on a piece of paper.

'But I really want to climb Everest,' I said. 'It's really import-ant to me. I want to use the opportunity to raise awareness about the planet.'

Lukas ran a company called Furtenbach Adventures, which had been leading mountain expeditions since 2014, though the team's experience stretched back over two decades. This was the organisation that my good friends, Fred and Marco, had signed up to do the climb with. When I told them I was interested in joining them, they couldn't give me Lukas's details quickly enough.

'He's the best there is,' Fred explained over the phone. 'Impeccable safety record. And he can get you up and down quicker than anyone else.'

Lukas, I quickly realised, was someone who did things differently. One of the biggest challenges of climbing Everest is acclimatising to the altitude. It can take at least a month to allow your body to adjust. As a CEO, I simply couldn't afford to take that sort of time out. Lukas's 'flash' expedition strategy took a different approach. Using the latest technology, he had devised an eight-week acclimatisation programme that you could do at home. This involved the use of a hypoxic tent, which prospective climbers slept in. Over the course of the programme, you were able to simulate altitude conditions right up to 8,000 metres. This reduced the acclimatisation time in the Himalayas and drastically shortened the length of the expedition – rather than two months, now the timeframe was down to around three weeks.

This wasn't the only difference about Lukas's approach. Most attempts on Everest begin in Nepal and take climbers up along the southern approach. Lukas starts his expeditions in Tibet, and takes climbers up the less well-known, and less climbed northern side. The northern route was steeped in history, which appealed to me, thanks to the climbs of George Mallory in 1921, 1922 and 1924. Secondly, Lukas believes the role of oxygen in the trip is absolutely crucial. Lack of oxygen was the single biggest cause of death on the mountain: by the end of 2017, out of the 288 people who had died on Everest, 168 of them had not used enough supplemental oxygen. The problem with oxygen was that it was both expensive and heavy to get up the mountain. The set-up that Lukas offered included two Sherpas per climber, guaranteeing the flow rate and amount of oxygen anyone needed. Rationing oxygen to a limited number of bottles might have made an expedition cheaper, but it came at a far greater risk: as so often in life, you get what you pay for.

Lukas was a young, good-looking Austrian. When he set up in business, he was called the '*enfant terrible* of expedition providers' by one newspaper. Everest expeditions had long been dominated by the same American and New Zealand companies – some with better safety reputations than others. When Lukas started, these companies were quick to criticise – people described his approach as 'snake oil', 'highly dubious and dangerous' and 'complete bloody hogwash'. But Lukas's results spoke for themselves. His model had the potential to become the future of climbing – an outlier for now, but the norm for tomorrow. As a businessman, I appreciated anyone who was prepared to think differently and shake things up. Lukas had a streak of seriousness to him as we spoke, but given I was asking to put my life in his hands, that didn't seem unreasonable.

'My company has an impeccable safety record in taking people up Everest,' Lukas told me. 'And we have an excellent success rate in getting people up and down the mountain. In 2017 and 2018, we had a 100 per cent success rate in getting people to the summit.' He paused. 'So it's not just anyone writing a cheque who I take on my expeditions. I have to be confident that they are capable of coping. Mountains can be extremely dangerous places.'

• • •

The nearest that I've come to dying in the mountains was in an avalanche in Alaska. I'd been staying in Valdez, a place that is famous for its heli-skiing. Some of its slopes are crazily steep, which is both a kick and a challenge. It also comes with risk – on the year when I visited, it felt as though someone was dying almost every day.

On this particular day, I was in a small group of experienced skiers, with the most amazing guide to take us down. He was the world freeride champion – an absolutely incredible skier. The helicopter dropped us off on an impossible-looking ridge that almost felt vertical. If I thought that mountain in India was steep, this one didn't even look skiable. Such was the slope that the only way down was to go straight with your momentum: there wasn't any other option.

Our guide was good, and he wanted to make sure he pushed us. To get to where he wanted us to ski, we had to traverse across the top of the ridge where we'd been set down. This was a super-steep section, peppered with rocks underneath. The guide went across first, showing us the route we needed to take. Having crossed, he turned back to give us a snow-gloved thumbs-up.

The rest of us followed. It felt a bit like walking across a tight-rope, such were the slopes down on either side, and I was as careful as I could be, taking it slowly, step by step. I was lucky that I was light, so didn't have much impact on the snow. The skier following behind me was actually smaller and lighter than me, but when he cut his skis a little bit too steep into the mountain, the impact was enough to pull the trigger.

The sensation was like having a rug pulled out from under your feet. One minute the ground beneath my feet was there, the next it had gone. I had barely enough time to jam my poles into the ground, but was grateful that I did. If I hadn't, I'd have followed the avalanche tumbling down. We clung on for dear life, hanging from our poles. There was nothing to do but to dangle and wait and hope.

I watched the avalanche unfold below. The snow tumbled down with a dark rumble, a low-frequency sonic bang. The landscape below disappeared beneath a cloud of white. I was too terrified to move – we all were, in case we triggered anything else. We all knew that we were lucky to be alive.

I'd only felt the full force of an avalanche once before, back in Turkey. I'd been skiing with Stephanie. At first I was confused, and thought she must have crashed into me at full speed. But such was the noise, the spray of snow and the smack of pain that I realised it was bigger than that. The wall of snow that landed on me was a bit like being hit by a truck. Before you know it, you're being shoved along and buried by it. That wasn't a large avalanche, but it was frightening enough.

Back in Alaska, I clung on to my poles and watched as the cloud of snow slowly started to settle. I glanced over at the guide, who was beckoning for us to inch over and join him. None of us wanted to ski after that, but there was no other way of getting back off the mountain. In a strange way, with the top layer of snow having been sliced off, it actually made skiing down that much easier. But my heart wasn't in it: I made it back down to Valdez as quickly as I could and hung my skis up for the day.

● ● ●

'It's your mental preparedness that's important here,' Lukas told me on our call. 'It won't be your physical ability that will be an issue with you getting up. Which doesn't mean there won't be a lot of work to do there. You're going to have to really put the hours in to get yourself into shape to get up. There'll be dietary stuff too, to get yourself ready.'

'I'll do it,' I said. 'Whatever you need me to do, I'll do it.'

From Lukas's response, I got the sense he heard that all the time. 'There's no shortcut here, Hakan. That's the first thing you have to understand.' He spoke into the phone with emphasis. 'Look, you're a practical guy, street smart. I'm sure you think that you can read everything, understand everything, get the right help here and there and get it done. But climbing Everest isn't something you can cut corners on.'

We talked on, about my family and work commitments. I was honest with him about that. 'My friend Fred is an entrepreneur, so he can be more flexible with his time,' I told him. 'I can't do that. But if you give me a schedule, I promise I'll stick to it.'

'I'm really worried about your lack of mountaineering experience,' Lukas continued. 'Some of the toughest parts of the climb are quite technical. To get up the north side, there are what is known as the Three Steps. They're all difficult, but the Second Step is famously difficult. There's a ladder hanging off the rock face to climb and you can't get it wrong: you're fully exposed, in a very hostile environment. There's a 2,000-metre drop straight down. If that's your first experience, you're going to struggle. There's so much to go wrong. You've got to make sure your boots don't get tangled in the ropes, that you don't rip your gear with your crampons. They're very easy to get stuck. And you'll be wearing these big mittens, will barely be able to hold anything. It's not a place for a novice.'

'OK,' I said. 'But what if I was able to practise that?'

'I don't know anywhere that's similar,' Lukas said.

'Maybe I could replicate it somehow,' I said. 'Build a model of it somewhere.'

'I've never heard of anyone doing that,' Lukas replied. 'But yes, in theory, that could help.'

'I make things,' I said. 'That's my job. Let me see what I can come up with.'

'Even if you do make something,' I could tell from Lukas's tone that he wasn't convinced, 'you'd still need some experience of climbing. You'd have to scale at the very least one substantial peak as part of that preparation.'

'I've thought about that,' I said. 'There is a group of colleagues at work who are going climbing in December. It's an annual thing, they go off and climb different mountains each year. This year, they're going to Argentina. Some peak beginning with A, I can't remember the name.'

'Aconcagua?' Lukas asked.

'That's the one,' I said. 'Do you know it?'

'It's the highest peak in the Americas,' Lukas said. 'Just under 7,000 metres tall.'

'Would that work?' I asked. 'If I could climb that, would Everest be possible?'

There was a long pause. Finally, Lukas replied. 'OK,' he said. 'Here's what we're going to do. I wouldn't normally accept someone under these circumstances, but I'm going to get you up Aconcagua. We'll build a training programme for you, get all the mountaineering gear sorted. And if you can get up Aconcagua OK, then we can look at pushing ahead and tackling Everest next year.'

'That's brilliant,' I said. 'You won't ...'

'I still can't guarantee that you will go up,' Lukas continued to be straight with me. 'All I can promise is that you will know whether you'll be able to go up or not when you get to Advanced Base Camp. So, if that's good for you, then we can proceed.'

Right from that first conversation, I trusted Lukas completely. I liked the fact that he was being honest with me – he wasn't just taking my money. From the way he sized me up, he obviously had a good instinct for people. It was now down to me to prove that I could do it.

'You've got a deal,' I said. 'I'll show you I can get up Aconcagua with no problem, and then we can take it on from there.'

How wrong could I be?

DEAD SEAS AND SKELETON COASTS

Stretching 500 kilometres between the town of Swakopmund and the border of Angola, the Namibian coastline offers one of the most intriguing and unusual landscapes on the planet. It is one of the few places on earth where the desert stretches to the sea. Tourist numbers are kept to less than a thousand a year in order to protect the land. Those that do go there can witness animals adapting to the harsh climate: elephants wading into the sea; lions, leopards and cheetahs patrolling the shore and sand for prey. But it is also a place where death is everywhere: the bones and carcasses of dead animals, but also the shipwrecks that have washed up over the centuries. Such is the danger for those navigating the seas here that Portuguese sailors nicknamed it 'the Gates of Hell'. As for the shoreline, the bones and shipwrecks gave another grisly nickname.

The Skeleton Coast.

More recently, this part of Namibia has achieved notoriety for a different reason. Those fortunate enough to experience this part of the world are noticing an additional feature to the local atmosphere: the smell. Today, the Skeleton Coast can reek of rotten eggs: the telltale sign of hydrogen sulfide in the air. Look out across the sea and you can see the water blanching white. On the shoreline, you can witness the ghoulish spectacle of thousands of dead fish washed onto the beach, their lifeless silver bodies glistening in the sun.

The water off Namibia was once one of the most productive set-ups in any ocean: a system of strong winds and currents helped to draw up nutrients from deeper water. This led to a rich harvest of plankton, which local shoals of sardines would feed on. But in the late twentieth century, the area suffered from overfishing, and stock levels never recovered. With fewer sardines to feed on them, the plankton began to gather in large piles on the ocean floor, rotting and decaying. This, in return, results in a mass of hydrogen sulfide and methane. When this rises to the surface, it turns the ocean white and suffocates the fish still living there. As temperatures rise, this process is accelerated: warmer water means more plankton growth, more rotting and decaying, and more hydrogen sulfide and methane.

For scientists, the appearance of hydrogen sulfide is a deadly harbinger. Rewind 250 million years, and the earth suffered the worst of its mass extinction events. In total, the planet has – so far – suffered five such occurrences, but the Permian–Triassic extinction was the worst, to the point that it is known simply as the Great Dying. The Great Dying was triggered by an enormous series of volcanic eruptions in what is now Siberia. Such were the scale and size of the eruptions that they released 100,000 billion tonnes of carbon into the atmosphere. Over the next 60,000 years, 96 per cent of all marine species and 75 per cent of land species died out. Unlike other mass extinction events, the Great Dying also took out the insect populations of the time. It took marine ecosystems 8 million years and the world's forests 10 million years to recover.

The volcanic explosions and the subsequent injection of carbon into the atmosphere led to higher temperatures on both land and sea. Sea surface temperatures at the equator reached 104 degrees Fahrenheit – a temperature at which no fish could survive. To a level, the oceans can soak up carbon in the atmosphere. But beyond this, the absorption of carbon dioxide leads to what is known as ocean acidification: essentially, CO_2 is converted into carbonic acid in seawater. A process of anoxification also took place – the oxygen in the water disappears, and this in turn suffocates the fish living there. During the Great Dying, the oceans lost three-quarters of their oxygen. A combination of this oxygen loss and the warming of the waters was the root cause of these mass extinctions.

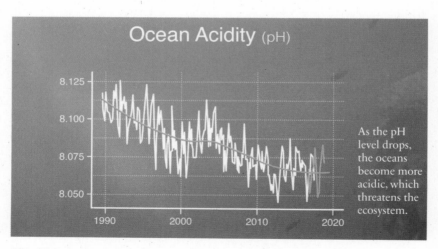

'World Scientists' Warning of a Climate Emergency', *BioScience*.
(Source: William J. Ripple, Christopher Wolf, Thomas M. Newsome, Phoebe Barnard, William R. Moomaw)

With the exception of the event that led to the demise of the dinosaurs (the Cretaceous–Paleogene extinction, 66 million years ago), hydrogen sulfide has played its part in every mass extinction. It starts by killing the animals in the ocean, then leaks out into the atmosphere and kills animals and plants on land. The sights and smells of Namibia's Skeleton Coast are a warning sign to us: that this is a process with the potential to pick up all over again.

• • •

'I think 2020 is the year when everyone who's got any ounce of sense has realised that the climate is going out of control. We are seeing that manifest through massive wildfires, not just in the US but Australia, Siberia, even beyond the Arctic Circle. And we're seeing it in terms of things happening in the ocean.'

Alex Rogers is one of the world's leading marine biologists. A visiting professor and senior research fellow at Oxford University, among those he advises on ocean ecology are the UN, Greenpeace, the WWF and the G8. He was also scientific consultant on the hugely influential *Blue Planet II* TV series and the author of *The Deep*, a fascinating look at the hidden wonders of the oceans and how we can protect them. So when I wanted to learn more about the effects that climate change is having on the oceans, there was no better place to start.

'The ocean is incredibly important in climate change for two reasons,' he told me, when we chatted over Zoom during the pandemic. 'The first is that, because of its size, it has a massive role in the planet's climate system. As a result of that, the ocean has already played a huge role in mitigating climate change. It's absorbed 93 per cent of the excess heat produced as a result of global warming. It's absorbed about one-third of the excess CO_2 that we've emitted. The ocean is now suffering the consequences of that.'

The role of the oceans in dealing with carbon cannot be understated. In those early models of the greenhouse gas effect, it was assumed that the seas would simply be able to soak up any carbon we produced. But those assumptions didn't factor in the amount of CO_2 we would go on to produce as the world population grew. And neither did they take into account what would happen to those creatures that lived in the waters as the sea temperatures started to rise.

In 2020, the world's oceans reached their hottest level since records began. Similar to the rise in global temperatures, the five hottest years in the oceans have taken place since 2015. Although firm data on ocean temperatures only goes back seventy years, the evidence suggests that the oceans are the hottest they have been for 1,000 years and are heating at their fastest rate for the last 2,000 years. Put simply, in terms of climate change, the warmer the water, the less it is able to soak up the excess carbon dioxide.

'The ocean is going to become less productive through a number of the effects of ocean warming, particularly stratification,' Alex explained. 'This is a reduction in the mixing of the upper ocean with deeper layers of the ocean, because of the warming of the surface. What that does is reduce nutrient supply to the surface, and that reduces primary production, and obviously from there production of fish and so on. Also, as a result of warming, fish are getting smaller. Which is a result of the physiological effects of that, of reduced oxygen concentration in seawater and so on.

'Associated with the absorption of CO_2 is ocean acidification. This is where CO_2 is converted to carbonic acid in seawater. That is also going to have some significant impacts for marine life. That will lead to the deoxygenation of the ocean as well, the simple fact that warm water holds less oxygen, and also because of this stratification effect of less ocean mixing.'

Warmer water has other consequences too. It results in more tropical storms, and makes them more severe. It has an effect on rainfall patterns, which in turn leads to floods, droughts and wildfires. It also results in rising sea levels and, as Alex explained, has a catastrophic effect on marine life.

'We're seeing, as a result of ocean warming, a global-scale redistribution of marine life,' he told me, 'and also an increasing level of risk to marine ecosystems. We're seeing fish stocks move away from the tropics and towards the poles. That means that a major source of protein and nutrition for many people in the developing world will essentially be decreased, and we're already seeing that manifesting in terms of shifting of fish stocks.

'The ocean is really being affected in many, many ways by climate change. The loss of some of these ecosystems, for example, means that human populations actually become more vulnerable to some of the other effects of climate change. If you lose coral reefs and mangrove forests and so on, you're losing coastal protection. And of course, other effects of climate change include increasingly ferocious extreme weather events: coastal protection becomes very important in defending coastal infrastructure from those types of events.

'Climate change is often talked about in terms of what's going to happen in the future,' Alex concluded. 'But we are already seeing really significant impacts on the ocean, and have been since the late 1970s, as a result of climate change.'

* * *

I've been fortunate, if that is the word, to see some of those changes for myself. I was a young boy when I first learned how to dive – I was about ten when my father taught me. The Mediterranean is a great place to learn, with the water being so clear to swim through. I did my training, got certified and then, when I moved to Hong Kong for work in my twenties, my love for diving really took off.

As I mentioned earlier, I was lucky to work for a chairman who was an avid diver. He had the contacts and resources to set up a number of amazing expeditions. Part of the fun would be the journey in getting there, island hopping by boat, small aircraft and sea plane. We travelled to places such as Raja Ampat, Kalimantan, Papua New Guinea, all kinds of wonderful locations. Some would be quite dangerous to get to, others so remote there wasn't really any local infrastructure.

The sort of things you saw blew your mind. There's something about being deep down in the water that can make you feel as though you're in outer space. The strange creatures you see are like aliens. On some trips we were lucky enough to identify new species – I remember we discovered a new type of squid on one trip, completely translucent, one of the most mesmerising things I'd ever seen.

But over the last twenty years, the pristine nature of even these more remote locations has ebbed away. Part of this is due to the rubbish and waste and plastics you now find everywhere (a subject we'll come back to later on). I remember going muck diving in the Lombok Strait; this is where you're searching in the water for the tiniest of creatures. It's one of the most exciting types of diving you can do, because the discoveries are some of the strangest and most alien creatures you'll ever see. But with the amount of rubbish in the water now, it's hard not to get entangled in plastics as you're diving. Elsewhere, I've seen species disappearing and declining. The number of sharks, for example, is continuing to shrink thanks to the illegal finning that some countries are involved in.

Turkey's Sea of Marmara, where I spent most of my child-hood, is also facing a similar threat. It's an inland sea that provides a rich and nutritious ecosystem for its inhabitants, as well as a source of life for those settled around it. You wouldn't think anyone can really kill a sea, but somehow, in the summer of 2021, Marmara came very close to dying.

Marmara is the most densely populated region in Turkey, and home to 24 million people. There are seven provinces that have coastlines on the Sea of Marmara. Due to poor judgement, weak infrastructure, and short-sighted development, it handles their waste in unimaginable quantities. Overfishing and pressure from waste intensified by global warming is destroying – if it has not already done so – a once beautiful, plentiful habitat. For more than three decades now, Marmara's health and wellbeing has been on the decline.

The now infamous 'sea snot' is not a new phenomenon. The first sightings in Marmara occurred in 2007, in deep waters. Nobody thought of interfering. By 2021, sea snot covered most of the Marmara's surface, endangering the life underneath – suffocating the fish and the coral reefs, poisoning the mussels and crabs.

With pollution and heat, certain types of algae in the sea overgrow to dangerous levels and block out the sun. The one we're seeing now across the surface of the sea is called phytoplankton, and it feeds on nitrogen and phosphorus, continuously present in Marmara waters due to poor waste management. Phytoplankton, under normal circumstances, works to regulate the waters and serve the wellbeing of the sea. But when it's faced with stress, and overgrowing, it releases a mucus-like substance, covering the surface, trapping underwater creatures in a giant furnace, cutting off their oxygen, entangling them and immobilising them until they die due to disease and starvation.

And then there are the coral reefs. 'In terms of biodiversity,' Alex had told me, 'these are the richest ecosystem in the ocean. Coral reefs host about a quarter of all fish species. Not to put too fine a point on it: they're dying as a result of excess heat, through this mass coral bleaching phenomenon. At 1.5°C of warming, 70 to 80 per cent of coral reefs will be gone. If we reach 2°C of warming, that would be probably more than 95 per cent gone.'

The bleaching that Alex refers to is one of the starkest effects of climate change you can witness in the waters. It takes place when the temperature of the water rises above its normal range over a sustained period of time. The way a coral system works is that it gets the vast majority of its energy needs through photosynthesis. This is done via a microscopic algae called zooxanthellae. At higher temperatures, the zooxanthellae go into a sort of photosynthetic overdrive, producing, among others, hydrogen peroxide (which you might know as the bleaching agent used in hair dye). To stop these chemicals damaging the coral still further, the coral rejects the algae, but in doing so lose all the healthy greens and browns they get from the regular photosynthesis process. To witness the change in corals, as they go from teeming with life to becoming a bleached, skeleton white, is horrifying.

In recent years, there have been a number of what are known as 'mass bleaching events'. In 1997 and 1998, such an event took place in the Indian and Pacific Oceans, killing 16 per cent of the world's coral reefs. A second took place in 2010, this time also affecting the Caribbean and Australia. A third, the most significant yet, took place from 2014–2017, affecting coral reefs across the world – around 70 per cent of the world's reefs were damaged as a result. The trigger here was an El Niño event: that, plus the rise in sea temperatures, was enough to push the corals beyond their thermal limit.

The loss of coral reefs is devastating in a number of ways. There is an effect on the communities that depend on them, both for food and tourism. But they're also a treasure trove when it comes to marine genetic resources: their role in drug research is surprisingly important. The first drug from the ocean came from a sponge living in the Caribbean coral reefs – this was developed into cytarabine, an anti-cancer drug. Essentially, because sponges can't move, their only form of defence against disease is of a chemical nature. In this instance, scientists found a mechanism in the sponges that stopped the DNA replicating in cells; scientists then used this to stop cancer cells from replicating. Cytarabine is not the only drug to have come from coral reefs – there are about a dozen on the market, with many others in pharmaceutical trials. Among other things, to lose these reefs as a resource is to lose a way of combatting disease.

· · ·

One person who is trying to do something about coral reefs is businessman and entrepreneur Scott Countryman. Scott, like me, grew up with a love for the ocean; when he wasn't fishing, he was diving; when he wasn't diving, he was surfing; when he wasn't surfing, he was sailing, looking for new places to try out his board. Scott runs the Coral Triangle Conservancy, a philanthropic foundation aimed at protecting coral reefs and marine ecosystems of the Philippine waters. The Philippines is one of 17 megadiverse countries which host 70–80 per cent of the world's biodiversity, but as Scott explained, the situation is becoming critical there: 'Right now the reef fishery population in the Philippines is at less than 10 per cent of their natural carrying capacity compared to 50 years ago; and less than 1 per cent of the remaining coral reefs are considered to be in pristine condition. Historically, the main causes of the decline of coral reefs have been overfishing, destructive fishing methods and land-based pollution. Many important food fish species have disappeared and coral reefs near areas of high population density have been pushed beyond the breaking point. In the past decade, with increasing coral bleaching and more extreme weather events caused by global warming, coral reefs have been dying (because of the past decade) at an alarming rate and the seed DNA we need to repopulate entire coral reef ecosystems is becoming increasingly scarce.' This has an effect not just on the marine life, but also on the local communities who depend on the sea for survival: 'We're talking about millions of people that rely on the oceans every day and tens of millions of people that indirectly rely on fish protein for food.'

Scott explained that the shallower waters are more vulnerable to the higher than average temperatures associated with global warming: 'the highest biodiversity and biomass are in the first 8–10 metres of depth, and these highly productive zones are getting hammered right now. In areas where surface waters are not mixing with currents and cooler deeper waters, the majority of hard and soft corals have already died in many locations. So you're losing the biodiversity of hundreds of different species of corals, and that cascades down to many thousands of species of animals that are dependent on each specific variety of coral. With fewer corals able to survive the above average temperatures, coral reefs are becoming less complex, which means less biodiversity, and more dominated by algae instead of soft and hard corals.

When this happens, the careful balance of the local ecosystem is disrupted and has cascading impacts on many other interdependent species. Fish can move away to cooler waters, but billions of other residents of each reef are not as mobile. This in turn affects the food chain. Creatures that might previously have been eaten can find the space to reproduce to unnatural numbers unencumbered without predators. Other species that might not have been present previously are now able to move in. Scott mentioned two species here in particular. Firstly, he talked about the crown-of-thorns, a particularly nasty type of starfish that crawls around on coral reefs and eats them. An average-sized adult (40 cm) can kill up to 478 square cm of live coral per day through its grazing activities. Unchecked by natural predators, which have been overfished to much lower populations, they can do even more damage to coral reefs than bleaching events. A pregnant female crown-of-thorns may contain anywhere from 12 to 24 million eggs, and may produce as many as 60 million eggs throughout a season. Nutrient additions from polluted rivers likely have contributed to outbreaks where crown-of-thorns larvae survival rates have greatly benefitted from nutrients. Thanks to a mixture of higher ocean temperatures and nutrient pollution from rivers ending up in the oceans, the number reaching maturity is now hundreds of times more than natural survival rates.

Another creature posing a challenge is the local jellyfish population. The island of Palawan has seen an explosion of blooms of bright pink tomato jellyfish *(Crambione cf. Mastigophora)* – these usually occur seasonally, and blooms are associated with seasonal changes in the ocean, like temperature or salinity. More blooms are happening now because of warmer waters in general. In the same way that the sea around Namibia's Skeleton Coast turns white, so the sea in the Philippines can now be pink in colour. Higher temperatures speed jellyfish production and extend the reproduction season: 'You drive the boat for an hour in one direction,' Scott explained, 'and there's not a square metre that didn't have at least ten all loaded with stinging tentacles. There were places where the wind was pushing up against the side of the shoreline and there'd be twenty feet of purple gelatinous goo. It looked like you could walk on them. Talk to the locals, and nobody can remember ever seeing these dramatically increased numbers of jellyfish.'

So what can be done to save the coral reefs? Scott spoke about needing to remain focused on reducing carbon in our atmosphere and solving the climate crisis. In parallel we need to set aside 30 per cent of coastal water of each country as ecosystem-sized coral and fish nurseries and ramp up efforts on innovation and technology to help marine life adapt to climate change. One suggestion is to genetically modify the coral reef to become more heat resistant. There is research being conducted on coral, where some species are able to resist temperatures above 40 degrees for short periods of time. If researchers can transfer this temperature-resistant symbiotic algae living in coral over, it may help repopulate the reefs. It's smart science, but it comes with a huge risk. 'You hear that, and you think, what could possibly go wrong?' Scott added. 'You just don't know at that microbial level what sort of damage that might cause. But that's how desperate the situation is.'

Another idea being discussed is to create 'arks of biodiversity' on land through cultured regenerative processes, essentially setting aside a small percentage of the thousands of enclosed coral reef atolls and creating controlled environments for entire coral reef ecosystems many square kilometres in size by tapping into deep cool waters outside of the lagoons. The idea would be that if we did work out how to lower the temperature of the oceans again, we'd have the species ready to repopulate them. It's a nice idea, but the timescale is enormous – to repopulate the degraded parts of the Great Barrier Reef could take hundreds, if not thousands of years. Equally, there are the sheer number of ocean species that are unknown – the amount of marine life documented is thought to be only about 10 per cent of the total. It's hard to create an underwater ark when you don't know about 90 per cent of the creatures you're meant to be saving.

In the immediate future, the answer seems to be hard work on the ground. Or rather, the water. One of the solutions, Scott thinks, is to find ways to protect the entire ecosystem in an area. So rather than trying to save dolphins or turtles or other individual species, the answer is to protect the complete ecosystem from ridge to reef. Rather than small pockets, this needs to be done in larger areas, hundreds of kilometres in size. Protecting everything has a sort of a multiplier effect, and when that happens, the ocean has a surprising capacity to bounce back to life.

This isn't a cheap thing to do. It requires buy-in from local communities: getting them involved in the management and development to gain a sense of ownership is important. Getting government help, Scott has found, can be difficult in countries such as the Philippines; private investment and funding from benefactors is currently an easier way of funding such projects.

• • •

Speaking to Scott and Alex, what was immediately clear was both their passion for the oceans and also just how serious they felt the present situation is. I was inspired by their energy and enthusiasm to find solutions, constantly searching for new and different ways to deal with the issues.

Alex told me about REV Ocean, a not-for-profit company for which he is Science Director. The company's aim is to make the ocean healthy again, with any profits made ploughed back into efforts to achieve that. At the heart of the project is a state-of-the-art research and expedition vessel, which is due to launch in a few years. The ship looks amazing from the pictures: part research vessel, part hybrid superyacht. It will have everything from over-the-side water sampling equipment to a submersible deep-sea robot and autonomous underwater vehicle, both capable of diving to 6,000 metres.

One of the aims, Alex explained, is to study the symptoms of climate change and how it's impacting the ocean, but also how the ocean could play its part in both greening our energy supplies and drawing down CO_2 from the atmosphere. This work is one of three ocean areas the company wants to research. The other two, coincidentally enough, are the focus of our next two environmental chapters; later on, we'll look at the issue of plastic pollution and the issue of overfishing.

CHALLENGE ACONCAGUA

I shut my eyes. It felt like I was drowning. I was struggling to breathe; my lungs felt thick and heavy, as though they were full of water. I felt trapped, unable to move, and I could feel my heart thumping ten to the dozen. It was as though as I was going under, waves washing over me, consuming me.

I snapped my eyes open. Pitch black. It's OK, I tried to tell myself, reminding myself of where I was. In a tent, the sharp whistle of the wind buffeting its canvas sides inside out. Beyond were the lower slopes of Aconcagua. I was in Argentina, out in the foothills of the national park. I wasn't drowning. Far from it: the rarefied air couldn't have been drier. Outside, the slopes were rocky and barren. There wasn't a drop in sight.

I felt completely out of it. My head was throbbing. I felt nauseous, like I might be about to throw up. And I was so tired. With the long flight from Turkey and travel across Argentina, I'd hardly slept for several days. Usually, I was a good sleeper. With my work and the endless travel involved, I'd mastered the art of being able to sleep anywhere. Put me on a plane, seat me in the back of a car, and I could be out like a light in seconds.

Here, though, I couldn't sleep. I wanted to, I was desperate to bank some rest. I knew that I had days of hard climbing ahead of me. Aconcagua was the mountain that I needed to prove myself on, but I'd learned from my training that it was going to be tough. If I couldn't get up here, there was no way I would ever convince Lukas to take me on his Everest expedition. And to have a chance of getting up here, I had to be in peak physical condition. Instead, here I was, my first night in the tent, struggling for breath.

It was terrifying. Every time I shut my eyes, it felt as though I was about to die. It was like being in a battle between two competing wills: part of me was pleading for rest, wanting to shut down for sleep; the other part was that age-old human instinct for survival kicking in, flooding my body with adrenaline to keep me awake.

I remember once reading an article about the interrogation technique of waterboarding. It sounded a similar sort of sensation to the one I was experiencing: the struggle to breathe, your nose and throat being blocked, swelling and gagging, the rising sense of panic. Every time I closed my eyes and tried to drift off, the sensation came back. All I could do as the wind howled past was to try and sit it out and wait for first light.

Making it through to morning felt challenging enough. The thought of then getting up and climbing one of the world's tallest mountains seemed laughable.

● ● ●

When I told my friends at work, who are all passionate climbers, that I wanted to join them on their latest expedition, they were delighted. Ever since 2011, a group of executives at Arçelik have been going on annual trips, focused around scaling a different mountain each time. In previous years, they had climbed Kilimanjaro, the highest mountain in Africa, and Elbrus in the Caucasus mountains, the highest mountain (depending on your definitions) in Europe. Both these mountains form part of what are known as the Seven Summits: the highest peaks in each of the continents, with the others being Denali in Alaska, Vinson Massif in Antarctica, Kosciuszko in Australia, Everest and Aconcagua. The term was the brainchild of American mountaineer Richard Bass, who became the first person to climb all seven when he summitted Everest in 1985. The definition of the seven remains in dispute – some climbers argue that Elbrus isn't really in Europe, and Mont Blanc should be included there. Equally, some suggest that Kosciuszko should be swapped for Puncak Jaya in Indonesia, to represent Oceania rather than Australia. But whichever list you go for, Aconcagua is firmly on every one, the second highest of those seven peaks, with only Everest higher.

At 22,830 feet, or 6,959 metres, Aconcagua is the highest point in both the western and southern hemispheres. Its name is thought to derive from the Quechua, *Akon-Kahuak*, meaning sentinel of stone. The mountain sits on the western edge of Argentina, about thirty kilometres from the border with Chile. Indeed, the history between the countries overlaps here: it was a Spanish governor of Chile, Don Garcia Hurtado de Mendoza, who ordered an expedition over the Andes in 1561, and the resulting city of Mendoza took his name; in 1817, General Don Jose de San Martin did the opposite trip, taking his Army of the Andes over the mountain passes in Chile, bringing to an end 300 years of Spanish rule.

All of this, of course, is the blink of an eye in the history of the mountain itself. Set slightly apart from the main Andes range, it is volcanic in structure and was active until about 9.5 million years ago, perhaps the reason for its extraordinary height. While human civilisation spread south from North America about 15,000 years ago, Aconcagua remained curiously absent from the myths and legends of region – possibly due to its inaccessibility and the difficulty in climbing it. However, in 1947, the remains of a guanaco (a close relative of the llama and alpaca) were discovered buried on the ridge between the mountain's two peaks. Assumed to be a sacrifice, the find was followed in 1982 by that of a 500-year-old Inca mummy, buried at 5,000 metres. Alongside the body of a young male were numerous items such as cooked beans, sandals and figurines, presumably supplies for the journey to the next world.

Whether or not the Incas ever reached the summit of Aconcagua, the present-day attempts to climb the peak began back in the late nineteenth century. In 1883, German mountaineer Paul Güssfeldt first attempted to summit, persuading locals to help with claims of possible treasure at the top. Bad weather and high winds hampered his attempts and he had to give up 500 metres short. Fourteen years later, an expedition by the English climber Edward Fitzgerald also attempted to reach the top. Fitzgerald was defeated by altitude sickness a few hundred metres short, but his Swiss guide, Matthias Zurbriggen, carried on and became the first known climber to reach the peak.

Now it was my turn. In the run-up to the trip, I had been feeling confident. After all, I was prepping for Everest, not Aconcagua. My training routine was going well, and I was feeling in good shape. I had the best gear mountaineering could offer; it was designed to survive the Himalayas, so I knew I'd be well kitted out here. The others in the climbing group were all older than me: the average age was early fifties, with one member being in their early sixties. If they could get up here, I felt that with my own fitness levels, it should be a straightforward rehearsal for the real climb the following year.

Such was my confidence that I joined the climbing party late. While everyone else set off on 22 December, to give themselves plenty of time to acclimatise, I decided to spend Christmas in Turkey and then fly on to join them. Christmas Eve is the main focus of the celebrations in my family, and I didn't want to miss that. My family all came together, and I enjoyed a farewell dinner.

I set off for Argentina the following morning; with time zones, I had an extended Christmas Day! The flight from Istanbul went via São Paulo and took eighteen hours. When I arrived in Buenos Aires, the airport was shut down, like a ghost town. I had a huge amount of climbing gear with me – fifty-one kilos – and needed a taxi to take me to where I was staying. By the time I found one, they were plucking rates out of the air and I was charged $25 for a two-kilometre ride. When I got to the Holiday Inn, I was then charged another $25 to change a $100 bill to pay the driver.

I hadn't slept on the plane, and didn't get much sleep at the Holiday Inn. After three hours, I was up again at 5.30 to go back to the airport and catch my flight to Mendoza. Here, I was met by one of our guides, Diego, who drove me into the city to pick up my climbing permit, and to get my gear checked. Once that was done, we got back in the car and set off for the national park, where the route to the summit started. This was a three-hour journey by car; the final section continued on foot, with your equipment being carried by mule.

The drive was both beautiful and deeply depressing. The landscape was sparse and stark, but also extraordinary in its emptiness: an arid, semi-desert setting without a single tree in sight, just this bleak red rockiness. It was like driving on the surface of Mars. As the road climbed, we started passing what used to be ski resorts. Ski lifts hung lifeless in the air, rusted and drooping, creaking in the wind. There were hotels with windows boarded up.

'There's no snow,' Diego explained. 'Not any more. The glaciers above, they're all going. And without the snow and ice, there's no water.' He waved his arm towards the dusty, rocky roadside. 'And so we get this.'

The lack of water, I learned, was an increasingly critical issue for the region. Over the previous decade, the region's glaciers were shrinking at a rate of twelve inches a year. The year after I visited, the two main rivers in the area (the Mendoza and Tunuyán) received just 10 per cent of the projected snow melt. Water was the main driver for the local economy: local agriculture and industry accounted for over 90 per cent of its use. Projections suggested that it wouldn't be long before the area was unable to meet its water demands. Crucial for the area was its role as a wine region. Mendoza had an international reputation for the quality of its Malbec, and produces 80 per cent of Argentina's wine. But the water shortage caused by global warming was beginning to affect its vines; this meant lower yields, sicker plants and a different taste to the wines. Some winemakers were migrating to higher slopes, others were simply giving up.

Seeing the effects of global warming first-hand was shocking. I'd read about how Elon Musk and others had plans to set up missions to Mars. Looking out the window, I couldn't help thinking that with the way we were treating the planet, wait a few years and you wouldn't need to go anywhere to imagine what visiting the red planet might be like.

The journey reminded me why I had decided to take part in this crazy scheme to scale Everest. I swore to redouble my efforts to succeed, to make the case about what was happening. That journey, I knew, began out here. I was looking forward to arriving, getting a good night's sleep, and setting off.

• • •

The morning after the night from hell, I looked like I'd been hit by a truck. My face was bright red and completely swollen. I was exhausted and dreaded the full day's walking ahead: before the climb to the summit, we had to undertake a number of practice hikes, to help our bodies get used to the higher altitudes. I felt so awful, however, I wasn't sure I could manage to make it beyond the camp.

I went to see the doctor. He took one look at me and went, 'Wow.' He proceeded to give me tests and check my oxygen saturation, but knew from the moment he saw me what the matter was.

'Altitude sickness,' he said. 'How long have you been here?'

'I got here yesterday,' I said, and talked him through my journey there.

'I thought your group had been here for days.'

'They have been. I, er, arrived late.'

Signs of altitude sickness.

'That's not a good idea,' the doctor said. 'Base camp here, we're at 4,000 metres. If you've come straight here from Buenos Aires, I'm not surprised you're feeling like that. Your body needs to acclimatise, to get used to the higher altitude. That takes days to do naturally. You've tried to do it in twenty-four hours.'

'OK,' I said. I felt as though I was learning an important lesson the hard way. 'So what can I do now? I really need to get some sleep.'

'I can see that,' the doctor said. 'Well, I can give you some drugs.'

He started telling me about Diamox. It was a drug a lot of American climbers use, and it helps to speed up the process of acclimatisation. 'It does have side effects,' the doctor said. 'It's a strong diuretic. You'll feel like you need to go to the loo the whole time. But at the same time, you'll be losing a lot of water, so will need to drink more than usual. Four or five a litres a day.'

'Right,' I said, not liking the sound of that. 'Anything else?'

The doctor nodded. 'You may find you get a tingling sensation in your arms.'

I'm not a fan of medication at the best of times, but particularly not a drug like this.

'Is there not any alternative treatment?' I asked. 'A more natural way of dealing with this sickness?'

'Oh yes,' the doctor nodded. 'There is a completely natural way of dealing with your situation.'

'I think I like the sound of that,' I nodded. 'So what is it?'

'I send you home,' the doctor replied. 'As soon as you head back down to Mendoza, you'll immediately start feeling a lot better?' He reached into the drawer of his desk and pulled out a packet of pills. He slid them across the desk to me. 'But if you want climb Aconcagua, you'd better start taking these.' The doctor gave me a look. 'I'll need to assess you again before you start the climb. If I don't think you're well enough to attempt it, I won't let you go.'

I did as I was told and took the tablets. As the doctor said, my arm started to tingle and I needed to go to the loo the whole time. I didn't enjoy feeling the effects of the drug coursing through my body, but faced with another night like the one I'd just had, I didn't have much of a choice. The experience stripped me of the over-confidence I'd turned up to base camp with: rather than joking with the others about how I'd been the smart one by waking up in my own bed on Christmas morning, now I realised they'd been the clever ones by coming early and letting their bodies adjust. And rather than boasting about my fitness regime and how I was going reach the summit first, my body felt listless and heavy on the practice hikes. I was being left behind, and could feel my body getting weaker as the walks went on. It didn't take long for my feet to be covered in blisters. Every step I took seemed to hurt.

Each morning, I went to see the doctor, hoping for the sign-off that I was OK, even though I was still feeling terrible. One morning, my face still red, I rubbed snow into my skin in the hope that it would cool it down and bring the swelling down. But the doctor wasn't fooled. To make matters worse, he then suggested I stop drinking coffee. I was really beginning to not like this guy! I can drink my body weight in caffeine over a normal day, so to have to cut that out, on top of everything else, was doubly hard.

But whether it was the drugs finally kicking in, my body adjusting over time, the lack of coffee or a combination of all three, my body suddenly clicked. I got an hour's uninterrupted sleep and almost immediately, felt a different person. From that point on, I started to get my energy back. I got some more sleep, was able to walk without problems, even my blisters seemed to calm down. I went to see the doctor again, who with a chuckle passed me fit to climb. I was ready to go.

* * *

Four a.m. Summit day. I was tired, but excited. Having acclimatised, I'd made up for lost time on the way up. At each camp, I was able to sleep for longer, which boosted my energy still further. I was tired for sure, but was up to the challenge.

From base camp, Plaza de Mulas, there are three camps on the way up to the summit: Canadá, Nido de Cóndores and Colera. The climbs each day were short but hard. They were what a mountaineer would describe as non-technical, so you were walking rather than climbing with specialised gear. But the surface was scratchy; there was a lot of scree which left you sliding and slithering your way up. And the combination of the lack of oxygen and the whip of the wind was killer. It was cold and exposed and as first light dawned that morning, the jagged silhouettes of the ridge looked rough and spectacular. Above, huge birds circled, which gave the place an eerie, almost prehistoric feel.

If the appearance felt timeless, there were elements of the morning routine that were positively space age. You weren't allowed to leave any waste on the mountain, and that included human waste. We were given bags to use when we needed to go: the same ones that NASA gives to astronauts in space. They have sand in and freeze solid once you've used them. Then the poor porters get the short straw of having to carry them down, a bag full of strange 'stones'.

A few days earlier, the raging wind and thinness of the air would have rattled me. But as the group made for the summit, I felt good and strong within myself. We hadn't gone far before I heard a commotion behind. One of our climbers was in a bad way. I scrambled back down to see if I could help. Being the oldest member of the group, he said he didn't want to slow us down or damage our collective chance of hitting the summit; so he decided to give up. It was disheartening to see that right before the summit push.

The summit journey from that final camp is a hard pull up. Those in the group who had climbed Kilimanjaro described it as being much more difficult. But the views looking down as we rose were spectacular. The stretch of the mountains beyond is stunning.

Just as we were about to reach the summit, I noticed another one of the climbers took a turn for the worse. His face was white and his lips were blue. I tried to speak to him but his replies didn't make sense. When I asked him questions, he often didn't respond at all.

On the way to the summit, Aconcagua.

'Oxygen deprivation,' said one of the guides, but the climber managed to get back on his feet and eventually continue to climb. I was messing around with the rest of the group who had made it, taking photos and videos. Someone joked that with Everest not being climbed at this time of year, we must have been the highest people standing on earth. A colleague shouted for us to huddle together for another photograph to celebrate. I bent down on my knees to pose for the picture and ...

Click.

I felt it instantly. My knee had gone. I couldn't believe it. I'd made it all the way up the highest mountain in the Americas without any real trouble. Now I thought about it, there had been a few occasions getting in and out of the tents when it had tweaked a little. Pulling my knee to put a boot on, it had felt a bit funny. But I'd always given it a bit of a wiggle and it had been OK.

On the summit, it was a completely different story. I knew immediately that this was of a different magnitude. My knee hadn't just loosened a bit. It had properly popped out. And as I crouched there, I wasn't sure how I was going to get it back in.

'Hakan, are you OK?' One of the guides came over. 'You're as white as a sheet.'

I shook my head and explained the situation. 'It's my knee. It's gone.'

Now it was the guide's turn to look alarmed. 'No,' he said. 'That can't happen. Not here. You've got to fix it.'

My mind flashed to thoughts of the climber Beck Weathers. I had read his book, *Left for Dead*, about his disastrous climb of Everest. On the highest slopes, it is every man for himself. You get an ankle injury and you're on your own. Beck was incredibly fortunate in that someone ignored that rule and helped him down. That guide, funnily enough, was now in charge of the organisation who had taken our group up Aconcagua.

I didn't want that to happen to me. I had to get back down under my own steam. I tried to pop my knee back in, but it didn't want to go. By now, I was beginning to panic. Even in the cold, I could feel myself starting to sweat.

'Try again,' said the guide. There was an edge to his voice, underlying how serious the situation was.

I took a deep breath. I wiggled it, I twisted it, I turned it any which way I could until finally, pop, thank God, I was able to push it back into place. Gingerly I stood back up.

'Can you walk around?' said the guide.

Slowly, I walked around the summit. I didn't want to put any weight on it at first, but I knew that I was going to have to if I was going to get down. My knee didn't hurt, which was a relief. But I could really feel it, and was worried it might go again at any moment. I was terrified about bending it. Could I really make my way back down Aconcagua without doing so?

'How does it feel?' the guide asked.

'OK,' I said. 'I think.'

'You've got to get down.' The guide eyeballed me. 'There's no choice in that. Do you understand me?'

I nodded. I couldn't believe how events had turned: one minute I was celebrating the climb, then in a split second I was worried about how I was going to make my way down. Like the altitude sickness at base camp, it felt another hard lesson learned. If you don't respect the mountain, it will take its revenge.

• • •

And Aconcagua wasn't finished with me yet. I took it easy going back down at first, descending and dodgy joints not being the greatest of combinations. But as my knee stayed together, my confidence started to recover.

I skidded my way down. The surface of the mountain, it turned out, might have been tough to climb up, but it was great for going down. That mixture of scree, pebbles and small rocks allowed you to slide your way down. Once you got the hang of it, you could go quite fast with the momentum – it was almost like skiing, except without the skis. I got into a rhythm of taking four or five running steps, then skiing and skidding down for twenty. Then another four steps, and so on. The technique kicked up a lot of dust – you wanted to be the person at the front doing this. But it was invigorating. I realised that reaching the top had taken more out of me than I had been expecting, and I wanted to get down quickly.

We got back down to base camp well ahead of schedule. And it turned out that was just as well. Behind us, above us, the clouds were coming in. It was the third warning sign that the mountain wanted to give me. We had barely made it back to base camp before the snow came in. Small flurries at first, and then big thick flakes, swirling around in all directions. Then the snow started to come in heavier, and the storm really hit. We'd booked for a helicopter to pick us up from base camp and take us back down, opting to skip the last couple of days walking down, but there was no way anything was moving in that weather. There was nothing to do but stay in our tents and wait it out.

Above us, still on the mountainside, were the porters, carrying all our stuff. They realised they weren't going to make it back in time, and so hunkered down on the mountainside. Thank goodness they had our gear with them. They used it to try and keep warm, but by the time they made it back, several of them had frostbite. That shocked me: these were meant to be the professionals, who knew the mountain trail inside and out, and yet here they were, caught in a snowstorm. I put my life in your hands over the last few days, I thought. That could have been me up on the slope.

When the storm cleared, I rearranged for the helicopter to arrive, and we flew back down. The snowfall reminded me of what the area should have looked like, before climate change had changed everything. We landed at one of the dead ski resorts, the rusting ski lifts creaking around above our heads, and decamped into one of the remaining open hotels. I was desperate for a hot shower, but more than that I wanted food. When I saw the breakfast buffet laid out, I knew I just needed to eat.

We sat there stuffing our faces, filling ourselves up with all the calories that we'd used up on the mountain. I realised how hollowed out I'd been by the experience, and how much it had taken out of me. But as my hunger became satisfied, so my sense of achievement grew. From a standing start, with no mountaineering experience, I had climbed Aconcagua. That felt good.

As we were talking, I noticed a young woman across the way going back and forth to the buffet. She was stick thin, in her early twenties, piling her plate up with a mountain of food and then going back for more. Where is she putting it all? I wondered. The next time she went up to the buffet, I joined her. I could see that her face was all cracked and burnt. Wind-whipped. I knew how that felt, and if it wasn't for a tube of cold-sore cream I'd had with me, my face would have looked the same.

'Here, let me give you something for that,' I said. I was feeling a bit full of myself for having completed the climb. 'You need it for where I've just been. Yeah, me and the guys over there. We've just bagged Aconcagua.'

'Oh, right,' the young woman looked nonplussed. 'Yeah, me too.'

I double-blinked at that. 'Wow, what, who with?'

'Oh, just me,' the young woman said.

'But who carried your stuff?' I asked.

'I did, of course.' The young woman looked confused. 'Didn't you?'

I thought about the team of porters, quite literally carrying my shit. 'So,' I changed the subject, 'how did you get back down in that blizzard?'

'Yeah, that wasn't so good, was it?' the young woman said. 'That was a hard walk back down from base camp.'

I decided not to tell her about the helicopter. 'Here,' I said, handing her my tube of cream. 'You should have this, I think you deserve it more than me.'

• • •

Aconcagua taught me a lot. The strength and might of a mountain. The danger of the weather. The vulnerability of the body. But it also taught me humility, too. I'd felt on top of the world for having been on top of the world, but that young woman had pulled me up sharp. So much so that when Stephanie texted, keen to know that I was all right, all I texted back was, 'It was OK.'

Such were the aches and pains of my body that when I boarded the plane back to Turkey, I was thinking, never again. Remembering the altitude sickness, I wondered whether I was cut out for this climbing lark. My knee bothered me as well: what if it went while I was on Everest? But then I remembered those ghost resorts, the way that climate change had scarred the region, and the reason behind why I'd started on this adventure in the first place came back to me. For all my doubts, I had to face them down and finish what I'd started. By the time the plane touched down in Istanbul, I had managed to convince myself to continue.

GHOST SHIPS
AND DARK FLEETS

It sounds like something from a crime thriller: a ship, washed up on the shore of a beautiful island, containing five unidentified dead bodies. But in December 2019, on the Japanese island of Sadogashima, that's exactly what happened.

Sadogashima lies in the Sea of Japan, fifty kilometres north-west off the coast of the Japanese mainland, Honshu. It is an island rich in history and natural resources. Over the years, it has served as a place of exile for emperors, Buddhist monks and creative artists. In the seventeenth and eighteenth centuries, it was home to one of the world's largest gold mine productions. Today, tourists visit to enjoy its mix of mountains, primeval forests, and its coastline of rugged shores, wild beaches and fishing villages.

In December 2019, the island received a different type of visitor. Washed up on its shore was the shipwreck of a wooden sailing vessel. On board, police found the bodies of five men and, separately, two human heads. Given the condition of the bodies, which were described in the local media as 'partially skeletonised', it was difficult to confirm whether the heads belonged to the bodies. All the indications were that whoever the men were, they had spent a long time out at sea. The only clue in terms of solving the mystery was the faded Korean lettering on the side of the boat.

On the opposite shoreline of the Sea of Japan sits the North Korean port of Chongjin. The third largest city in North Korea, Chongjin was known for many years as the City of Iron, due to the heavy industry in the region. But today, the city has a different nickname: Widows Town, to mark the growing number of fishermen's wives whose husbands set out to sea and never return home again. The shipwreck that washed up on Sadogashima in December 2019 was far from an isolated incident. All along Japan's western coastline, the grisly phenomenon of 'ghost ships' can be seen; there have been over 600 boats so far, and counting.

But why are the bodies of these Korean fishermen being found so far from home, and in wooden, rickety boats that are in no fit state for such a long journey? The answer is, on the surface, economic, and underneath, environmental. Desperate for money following the collapse of its economy, the North Koreans sold rights to their fishing waters to the Chinese, for an estimated $75 million in hard currency each year. In 2017, additional UN sanctions on North Korea because of its nuclear programme stopped this transaction. But despite it being illegal, the dominance of the Chinese boats in Korean waters has continued. A 2020 report by Global Fishing Watch suggested that 900 Chinese ships fished in Korean waters in 2017 and 700 in 2018. Their haul for these two years is estimated at over 160,000 metric tonnes of squid, worth £346 million.

These Chinese ships are, ironically, known as 'dark fleets'. Ironic, because the ships fishing for squid are bathed in light. The plankton and fish species that the squid feed on are attracted by light. The squid follow their prey, where they then are caught using jigging lines. The lamps are strong: such boats routinely have over 100 lamps at 3,000 kilowatts each. Such is the brightness that fleets of squid ships appear on satellite images – small cities in the middle of the sea.

Squid is increasingly valuable as a catch on the international market, and not just in Asia – the US imports 80,000 tonnes a year, mainly from China. That value rises as stocks fall: the South Korean and Japanese squid stocks have declined by 80 per cent over the last two decades. All of which serves to explain why the Chinese dark fleets are in Korean waters, and why the displaced Korean fishermen are risking all in travelling further from home, into Russian and Japanese waters to illegally fish themselves, and paying the ultimate price.

• • •

In the 2020 Brexit negotiations over the UK's withdrawal from the European Union, the issue of fishing was one of the most vexed, to the point that disagreements threatened to scupper a deal, despite the fact that the value of fishing to the UK economy amounts to 0.1 per cent of British GDP. For supporters of Brexit, fishing was symbolic of their 'take back control' message, the romantic image of small trawlers fishing in the nation's waters encapsulating what reclaiming sovereignty was all about: 'We've got our fish back,' claimed Leader of the House of Commons Jacob Rees-Mogg. 'They're now British fish and they're better and happier fish for it.'

The mental health of a nation's fish aside, a quick study of the British fishing industry paints a different picture to the one of the plucky trawler often portrayed in the media. In 2018, an unearthed investigation revealed that more than two-thirds of the UK's fishing quota is owned by just twenty-five businesses, and almost a third is owned by five families on the *Sunday Times* Rich List of the 1,000 wealthiest individuals and families in the UK. The biggest share of the quota is not even owned by a British company at all, but the subsidiary of Cornelis Vrolijk, a Dutch multinational.

Modern-day fishing is far removed from the idealised image that many of us have from seaside holidays and fishing villages we may have visited. It is big business, industrialised, uses techniques that would make you blanche and is often highly illegal. The discussions over Brexit, although often heated, were in some ways fairly tame. This was a discussion over rights within a nation's fishing waters. But most of our seas and oceans fall beyond one country's jurisdiction. Here, there remains more of a fishing free-for-all. The resulting depletion of the world's fish stocks through this industrialised overfishing is close to reaching a point of no return.

Although a number of countries are involved in illegal dark fleet activities, it is China that has by far the largest number of ships. Research by the Overseas Development Institute (ODI) claims that the Chinese have almost 17,000 vessels involved. Again, the majority of these ships are thought to be owned by a few companies, who are able to operate through a series of subsidies by the Chinese government. Tax exemptions given to these companies, predominantly on fuel, are said to be worth $16.6 billion.

'The coast of China has been absolutely wiped out,' Scott Countryman of Coral Triangle Conservancy, explained to me. 'The fisheries there collapsed a decade ago, from the pollution from their rivers. It's an absolute catastrophe what has happened there. So their fishing fleets are now all over the world. Argentina just sank a boat off their coast. Down in Antarctica, they're going after everything from the brine shrimp to the krill, to pollock, haddock and deep-sea fish. These boats are factory boats, they stay out for years at a time. Some of them are 300 metres long and they can pull in 20,000 tonnes of fish in one net set.'

Scott told me about some of the techniques used: 'They use the most destructive fishing techniques that have horrific by-catch. Sharks, rays and large marine mammal megafauna all get caught in these nets. Some of these long lines draw kilometres long with hundreds of thousands of hooks. You've got pursing nets that are just wrapping up everything. The bottom of the South China Sea looks like every square inch has been dragged with trawling nets.'

Not only do these fleets fish in international waters, but they move in on other countries' fishing zones too. North Korea is far from an isolated example. With minimal regulation and often even less enforcement, these illegal fishing operations are able to operate with impunity off much of the coast of Africa, and places like the Philippines. They sweep in, clear out the local fish stocks and move on.

• • •

One person who knows how hard it is to stop these fleets is Pete Bethune. Pete is one of those guys you're glad to have on your side: an all-action environmental hero who claims that 'you haven't lived until you find a cause worth dying for.'

Pete's career began as a wireline engineer, studying reservoirs to find the best places to extract oil. But his work in this sector soon led to a growing interest in the role of energy and alternative fuels in particular. Studying for an MBA, he looked at how biofuels could be used in transport. This led to a realisation about the environmental situation we faced. 'All recoverable oils will be gone in fifty years, and we will just continue to deplete the earth's resources,' he said in an interview in 2019. 'When you're inside the industry, those facts start to eat away at you.'

Pete got out of the industry and began championing the cause of biofuels. He built a biofuelled powerboat, *Earthrace*, with the aim of breaking the world record for circumnavigating the globe, in order to raise awareness and spread the word about biofuels (he shaved two weeks off the previous record).

While on his record-breaking trip, Pete became aware of the damage being done to our oceans. He went diving off a Portuguese island and was shocked at the emptiness of a sea cleaned out of fish. In Fiji, taking shelter from an incoming cyclone, he found himself joined by a fleet of Chinese ships, all of which had been fishing illegally in the waters. Rather than returning to the day job after his round-the-world attempt, Pete instead decided to turn his attention to protecting the planet, and to the issues of marine conservation and overfishing in particular.

When I spoke with Pete in late 2020, he was in Costa Rica, trying to deal with the increasing encroachment of Chinese fishing in the area. 'There's a beltway,' he explained to me, 'a massive marine highway between Malpelo, Galapagos and Cocos Island. There are a lot of pelagic species moving between those three. And the Chinese are getting very clever about this.'

The numbers of ships are huge. In summer 2020, the Ecuadorean Navy spotted a total of 340 ships in the Chinese fleet. The biodiversity found in and around the Galapagos Islands is one of the rarest and richest in the world – it was the inspiration for Charles Darwin's theory of evolution. The aims of the Chinese fleet were somewhat less lofty. As Pete describes it, part of the battle is technological: fishing ships are meant to keep their AIS systems switched on, so that they can be tracked. The Chinese fleet swap between switching theirs off and undertaking a practice known as 'spoofing', where their GPS sends out a fake signal, suggesting the ship is somewhere else. This then allows them to slip into Ecuadorian and Costa Rican waters undetected. 'They know exactly where they are,' Pete says, in case anyone suggests they stray into these waters by mistake. 'They have big helicopters spotting the fish, and then use enormous purse seines (sweeping circular net 'curtains' which draw together to capture the fish). It's industrial fishing on a massive scale.'

Pete is trying to fight technology with technology. He is repurposing a US Navy ship to help patrol the waters. Through sponsorship from firms such as Schiebel, he has been able to acquire a drone to help detect the ships. The drone wasn't cheap: it cost $4 million, with a further $500,000 needed for training and spare parts. Satellite imagery is also vital. Although the ships can switch off their GPS equipment, their lights can shine bright enough to be picked up by weather satellites.

Pete described a previous mission in the Philippines where he used this satellite technology to help spot and stop a small fleet of boats illegally fishing. 'They were fishing in tribal waters where indigenous people have fishing rights. We spotted this fleet using the US National Oceanic and Atmosphere Administration satellites. We went out and caught these fifteen boats. Three of them were purse seiners: all of the rest were facilitating boats. They had light boats, with these huge lights on them throughout the night. They were taking twenty-five tonnes a day.'

Finding the ships, however, is just the first part of the process. Pete knows from personal experience that you can't just board the illegal boat: if you board a ship without the captain's permission, then you're in breach of international and maritime law. On an anti-whaling expedition in the Antarctic in 2010, Pete boarded one of the Japanese ships involved, and ended up being arrested and convicted back in Japan (having spent several months in jail awaiting trial, he was given a two-year suspended sentence and deported back to his native New Zealand).

These days, Pete makes sure that there is someone on board who has the required authority to intervene. In the Philippines case, he had a local tribe elder on board, which meant they could enforce the fishing laws. But this doesn't always happen: in poorer countries, there is often a fear of retribution if they make arrests. Pete talks about an 'underbelly of criminality' to some of these fishing operations. Often, it can feel easier for the local authorities to turn a blind eye.

Pete described the two biggest challenges of tackling over-fishing as a lack of resources and a lack of international agreements. On the first point, most of the conservation groups carrying out this work are of a smaller scale: Pete has his kitted-out boat; another similar organisation, Sea Shepherd, has fifteen. And although Pete travels the world to offer training, even once navies know what to do, they often don't have the resources to do it: 'To give you an example, Palau has created a marine reserve. There's no industrial fishing allowed in any Palauan waters. But Palau is a very small nation. They're a US protectorate, so they have a little bit of resources coming from the States. But the waters are enormous: to patrol that needs heavy, heavy investment and Palau just doesn't have the money.'

Even when countries do have the resources, patrolling for illegal fishing is not always a priority. Pete described how New Zealand suffers from such fishing in its northern waters: 'We occasionally send out an Orion aeroplane to do surveillance. We send a navy ship out once a year.' Part of this is down to a lack of awareness over the issue, and part is about cost: 'For you to send a ship 300 miles offshore and arrest a boat, that's a $30,000 exercise. For you to have a ship patrolling all year, you're probably talking

$10 million. For small countries, that's a hard ask.' And even when that money can be found, it still remains a fraction of the budget the illegal fishing fleets are working from – remember that figure from earlier about how the Chinese fishing industry is subsidised to the tune of $16.6 billion.

The second issue Pete cites is the lack of international agreements that are enforced. 'There is an agreement over the Patagonian toothfish or Chilean bass,' Pete gives as an example. 'This is found in sub-Antarctic and Antarctic waters. There's an agreement that a number of countries signed up to, which is controlled by New Zealand and Australia. If Spanish or Portuguese boats want to go and fish there, they can go and get a quota issued to them. The problem is that New Zealand and Australia, theoretically, have no jurisdiction in sub-Antarctic waters. So while some countries have signed up, like Portugal and Spain, China is like, "Fuck that. We're just going to go down there." And so, what mechanism does New Zealand and Australia have to enforce it? None at all. You're up to the goodwill of the other nations to abide by the rules.'

In international waters the situation is even muddier. Pete gives the example of yellowfin tuna, which is listed as being overfished in both the eastern Pacific and Indian Ocean, where stocks are at a critical level. 'The area between the Marshall Islands and America is enormously rich for tuna. But how do you stop a Chinese boat going and fishing there? No one can legally board that Chinese boat without the Chinese government's permission. And even if you are going to board it, where are you going to take it to court? At the moment, it's a free-for-all.'

. . .

If issues of resources and regulation can be overcome, then the creation of marine conservation zones can also offer a way forward, both to tackle the challenge of overfishing and also to protect marine systems and encourage them to recover. One project I know well is in Gökova Bay off the Turkish coast, run by the Mediterranean Conservation Society. In 2017, the project won the Whitley Gold Award, known as the 'Green Oscars'.

The president of the society is conservationist Zafer Kizilkaya, who has been running the organisation since 2012. A few years before, Zafer had been involved in the rehabilitation of a stranded Mediterranean monk seal. When the pup was released in the bay, Zafer dived down, and was horrified by what he saw. 'Underwater was a kind of nuclear war,' he remembered. 'There was no algae, no fish, just empty racks and discarded fishing gear.'

Zafer started researching fish levels in the Mediterranean and discovered that Gökova was one of the worst areas in terms of fish stocks, with only 4 grams per square metre (that contrasted with protected areas off the coast of Spain and France, where the level was 120 grams per square metre). This wasn't just bad for the Mediterranean, but also for the communities who relied on it: the catch of local fishermen was down by 60 per cent and they were struggling to survive.

Zafer explained how such a situation can create tension between the fishing community and conservationists. Was there a way of helping the former while protecting the sea at the same time? The starting point was to set up a marine protection area in the bay. With the launch of the Mediterranean Conservation Society, Zafer had both the authority to set up no-fishing areas and also the responsibility to enforce this. The no-fishing areas total twenty-seven square kilometres, with Zafer and his team monitoring fish stocks beyond this, in conjunction with the local fishing community.

As with some of the stories described elsewhere, illegal fishing is not an immediate priority for the Turkish Navy: it is down to Zafer and his team to patrol the bay. He currently has a team of six marine rangers, though in an ideal world it would require double that number. The marine rangers are linked back to the headquarters on land, and are equipped with duty cameras to log anyone fishing illegally. These are then reported to the Ministry of Agriculture and the coastguard to deal with. The rangers react to reports and sightings and vary their routines so no one knows what time they'll be patrolling. They also have a member of the coastguard on board from time to time – again, so those trying to break the law never quite know who they'll be stopped by.

The mark of success in this sort of project, Zafer explains, is when you see the return of apex predators: the largest fish in the food chain. If the system is healthy, and the stocks are recovering, then these will be in evidence. If the stocks dwindle, then the apex predators look for food elsewhere. Over the course of the project, Zafer has watched the return of sandbar sharks and dusky groupers, while the monk seal, whose numbers had dwindled to a solitary pair, is now back up to nine and growing. Once all but depleted, the 'no-take' zones are now teeming with fish again.

As well as the depleted stocks from illegal and overfishing, the other challenge that both Zafer and the local community has faced is how the marine life in the Mediterranean has changed as a result of the climate crisis. As water temperatures have increased, a number of invasive species have travelled up the Suez Canal and into the sea. Here's where the apex predators are so important: if they're in evidence, then they can stop the invasive species from growing in number. Without them, their populations can grow uncontrollably.

Part of the challenge is persuading local fishermen to catch and sell these new species, to also help keep these numbers down. To help convince the local community, Zafer's team have worked with chefs to come up with recipes and set up campaigns to persuade people to buy and try these different species. Other invasive fish have been caught for different reasons. One of the species that has arrived in the Mediterranean is a puffer fish. This fish is toxic – it has a neurotoxin called tetrodotoxin, which is 2,000 times more toxic than cyanide. Here, rather than being sold for food, the fish has the potential to be used for medicine: a pharmaceutical company from the US is experimenting with using the toxin in a painkiller they are developing.

The Mediterranean Conservation Society project in Gökova Bay is a model that can be replicated and used elsewhere. It shows how a balance of enforced marine conservation areas can bring fish stocks back, and balance these conservation needs with those of the local community. The more protected areas there are, the more they have the potential to connect up, and this in turn creates a multiplier effect in terms of conservation.

How large a scale would the projects need to be replicated on? Zafer estimates that 'to see biodiversity and sustainability of oceans, we need 30 per cent of them protected by 2030.' That would require a step-change in current funding. 'Annually, world fisheries get $35 billion to catch more fish every year,' Zafer reminds me. 'Every year more people are catching more fish with more crazy technology. To protect 30 per cent of the world's oceans, we need $100 billion. We need to change our mindsets. Catching more fish is not the answer. Protecting more is the future.'

• • •

In the meantime, we can all be more conscious and aware of what fish we buy and eat as consumers. According to Scott Countryman, as an example, between 20 per cent and 32 per cent ($1.3–2.1 billion) of wild-caught seafood US imports are illegal, and in other countries the figures are much higher. Knowing what you are eating, and where it comes from, is important. A lot of fish that you find in fast-food restaurants or on supermarket shelves has been caught on the factory ships where complicated supply chains foster illegal seafood, especially in reprocessing.

Knowing the source of the fish you eat isn't easy. 'A fish that is now being caught by the Chinese can be sold to a Malaysian company to transport it, is then packaged by a Thai company, goes through a sanctioned quota elsewhere and gets mislabelled as something else.' Scott talked about a case of DNA testing of fish-cakes of Manila, which were found to contain various threatened rays, endangered sharks and high overfished marine animal DNA. He also mentioned a 2017 investigation into fish restaurants in San Francisco, which revealed that fish on the menu was often mislabelled. Four restaurants claiming to sell red snapper were instead offering customers yellowtail rockfish, silvergray rockfish, white bass and Japanese sea bream instead.

If you're not sure where the fish is from and can't trace it back to where it was caught, you probably shouldn't be eating it. And think, too, about how much fish you eat as well, if at all. 'The number one thing we can all do as consumers,' Scott said, 'is to dramatically reduce your fish consumption, or just stop eating fish.' Scott cited Sylvia Earle, the legendary marine biologist, who gave up eating fish forty years ago. A few years ago, we were lucky enough to have Sylvia speak at a conference on wellness and the environment at Kaplankaya, so I heard her eloquent thoughts on fish myself. 'Think of them as wildlife, first and foremost,' she said in an interview in 2014. 'I have come to understand the value of fish alive in the ocean, just as we've come to understand the value of birds alive to keep the planet functioning in our favour. Imagine a world without birds. Imagine a world without fish.'

PREP, PRACTICE AND PERMISSION

The first thing Stephanie did when I told her I was going to climb Mount Everest was to throw her glass of wine at me.

On my return from Aconcagua, I had insisted on whisking Stephanie away for a night to make up for being away. I booked a room at a hotel in Istanbul and made a reservation at a restaurant that a friend who knows about these things had recommended. It was a very intimate place for dinner, and the few tables that were there were closely packed together. My plan had been to have a romantic evening, then tell her about my plans the following morning. But in the moment, as we sat in the restaurant, I couldn't contain my excitement and decided to tell her there and then.

I have to confess that when I first decided to climb Everest, I kept my plans extremely close to my chest. I discussed it with Lukas, and I told my friends Fred and Marco, because they were already committed to going, but I swore them to secrecy. And so as far as Stephanie, my family and my company were concerned,

I was going up Aconcagua on a work expedition and that was it. Everyone was impressed with how much effort I was putting in for the training, but they assumed that was just me being me in going the extra mile. I didn't want anyone knowing what my real plans were, because I had doubts about whether I could pull them off. If I didn't tell anyone, then there would be no humiliation when I pulled out.

Now I was back and I knew I could attempt Everest, I had to start telling the people that mattered that I was going to go. I didn't have long to do that. I arrived back in Turkey on 4 January. The deadline to pay the money to go on the trip was six days later. I had to talk to Stephanie, and then with work, for them to agree to give me the time off. Neither conversation, I knew, would be easy.

Stephanie's immediate reaction was one of shock and surprise. She felt set up, which I understood, even if I tried to protest that I hadn't planned to tell her at the meal.

'Whatever.' She signalled to the waiter to pour her another glass of wine. 'I can't believe you'd hide something like this from me,' she reasoned.

'I didn't,' I tried to counter. 'I didn't know for sure this was going to happen.'

Stephanie knew me better than anyone else. She knew how much I needed the wisdom and the warmth of her support. Once the shock of the news died down, and I'd talked through the reasons for my making the trip, I could see she understood why I was committed to going.

'How dangerous is this?' she asked. 'Yes, you've made it up Aconcagua, but you're not a climber. And you've got children. I don't want to end up raising them without you.'

I tried to reassure her about Lukas. 'He's the best,' I said. 'His safety record is second to none. Nothing's going to happen to me.'

But even as I said that, memories of Aconcagua flashed back. And Stephanie, I could tell, still felt concerned for my safety. In the end, I had Lukas to thank for reassuring her.

Lukas and Stephanie went ski touring in Innsbruck, near my daughter Andrea's ski camp. I didn't say who he was, that he was the owner of the company or anything: I just told her that he was an amazing skier. (When Lukas sent a message to Stephanie, one of our good friends saw his picture. 'Did Hakan arrange for you to go skiing with this guy? He's super good-looking.')

Stephanie is a great skier, but Lukas put her through her paces. He's the sort of winter sports person who flies across snow, moving effortlessly through the landscape and barely leaving any tracks. Stephanie was in awe of him as she followed Lukas up and down the mountain. As the two talked, and Stephanie worked out who he was, she told him about her concerns about the trip. Lukas explained about how he had two small kids himself, and that he wouldn't put me or him in any danger on the expedition.

'You can go,' Stephanie said, when she returned from the day. 'Lukas has promised me he'll bring you back down, no matter what happens. Having seen him ski, I think he could probably carry you on his back if he needed.'

• • •

Next, I had to speak with key executives at Koç Holding, our majority shareholder. It was one thing to take time off over Christmas on an expedition involving various members of the company. It was another to take time out to climb Everest. With the climbing window to scale Everest being relatively short, the time to climb didn't coincide with any holidays. And even with Lukas's ground-breaking hypoxic training technique to reduce the time the trip would take, it would still mean I would be out of the office for at least three weeks. And that was if everything went right.

I knew if I went straight to see the management team, they would turn me down. Fatih, who I still report to today, and I, had experienced Aconcagua together. We have a strong bond of confidence – enough to trust each other with anything. He knew a lot about mountaineering, and even he thought my chances were slim. When I told him about my plan, he shook his head and said, 'there's no way it will happen. There's no way they'll allow you to go. It's a nice dream, Hakan, but nothing more.'

'I'll give it a shot,' I said.

I went to see Arçelik, Rahmi M. Koç, who is also honorary chairman of Koç Holding, our main shareholder. He's an amazing man, now into his nineties. He possesses an unending life energy. His memory and intelligence is exceptional to say the least. He is someone I've had a very strong relationship with all my life. I asked if I could see him at his house, and he suggested we have breakfast together. At least I won't get any wine thrown at me this time, I thought.

The chairman sat there politely and listened to my arguments. 'Look,' I said, 'business is doing really well at the moment. It's true you won't have a CEO for a few weeks, but we've got a great team here: I trust them to take care of things while I'm gone. I'm sure they'll do a good job.' Not too good, I thought to myself, or I won't have a job to go back to! 'And I think the expedition will do a lot of good for the company. In terms of branding, in terms of my leadership, it plays really well. I really want to do this,' I concluded. 'I'm ready.'

The chairman didn't reply straight away, but mulled it over. I took a sip of coffee while I waited and tried not to look nervous.

'How risky is it?' he finally asked.

'It's risky, yes.' I held my hands up. 'I'm not going to lie to you about that. This is Everest, after all. But I'm going with the best outfit possible. They're incredibly experienced. I couldn't be in safer hands.'

'Hmm.' The chairman didn't sound convinced. 'You know, I don't feel comfortable about this, about you going. But at the same time, I don't think it right for me to stop you doing something that you really want so much.' The chairman raised a hand to stop me from interrupting. 'Other people have had sabbaticals, they've gone here and there, but I've never granted a request like this before.' The chairman looked straight at me. 'You're going to go and stand where Hillary stood? Sir Edmund Hillary?'

I nodded. The chair gave a short shake of the head. 'Well, if you do manage to pull this off, you will be different for the rest of your life.' He breathed out a sigh. 'Just promise me two things.'

'Of course,' I said. 'Anything.'

'Promise me that you will be very careful, and that at the slightest risk you'll turn back.'

'Absolutely.'

'You say that now,' the chairman said, 'but I know you. Once you're on that mountain, the hardest thing you can do is to turn back. There'll be that voice in your head telling you to go on, that it's not much further. But such a decision could be fatal.'

'I understand,' I said. And I did, twice over. My desire to succeed was a useful skill in business; on the mountainside, it might be my undoing.

'And I presume you'll take our flag, put it up there?'

I nodded. 'I'm sure I can find space in my backpack for that.'

The chairman held out his hand and I shook it. We had a deal.

That was the key moment. Having got our chairman's backing, I was able to go to Levent, who runs Koç, the holding company. I have worked with him for a long time and have always trusted his guidance and insights. He is a leader who believes in the higher purpose of the people he works with and supports their passion. After a brief but pleasant discussion where he stated his genuine concerns, mainly around my personal safety, he neither discouraged nor encouraged the expedition, as it was going to be on my own time and in a personal capacity. One highlight from this encounter was that he clearly understood why I wanted to embark on such a dangerous undertaking.

There was one last person I needed to speak to. My father. I sent him an email first, explaining my plans, then paid him a visit. 'I'm going to do this,' I said, 'but I'd really like your permission to go.'

My father gave me a look. 'You've already decided,' he said. 'You've already got everybody else's permission. You're not asking me, you're telling me you're doing it.'

I winced. As usual, my father was incredibly perceptive.

'Not only that,' he continued, 'but I'm the last person to know.' My father was a good friend of the chairman, Rahmi Koç. 'How does he get to know before me?'

In the same way that I needed the honorary chairman to persuade the company, so I also needed the honorary chairman to persuade my father. I was nervous about asking him, so knew that if others had already agreed to my going, my father was less likely to refuse. So yes, I was asking him last, but at the same time, it was a mark that I respected him the most.

My father, like any parent, was worried about his child doing something dangerous, but he nodded his acceptance. It had been a week of tricky negotiation, but I had done it. There was no going back now. I was committed.

• • •

It was still dark in the house when I slipped out of bed, leaving Stephanie sleeping. In the kitchen, the clock on the oven said 5.30. It was time for breakfast in my new routine.

I reached for the bottle of olive oil, glugged an amount into a glass. Then in the vegetable drawer, I found a bulb of garlic. I pulled off a clove, then using a knife, nicked off the end before sliding the blade underneath the skin to peel it off. With a garlic press, I crushed the clove into the glass of olive oil, gave it a quick swill and a stir. In the kitchen light, the oil glistened with a touch of gold. As I looked down at the glass in my hand, I felt my stomach twinge in anticipation.

If you want to climb Everest … a voice in my head said.

I winced, shut my eyes and downed the drink, trying not to gag as it went down.

My morning shot, though, was just the starter. In the juicer, I chopped up and chucked in a bag full of fresh green vegetables. In went a load of ginger and a spoonful of turmeric for good measure. Clamping the lid down, I held on to the juicer as it whizzed and blended the vegetables for thirty seconds. When it had finished, I poured its contents into a second glass. While my morning shot had glistened and refracted the light, the vegetable smoothie was murky and viscous, a dark swampy green. I could still taste the raw garlic in my throat. This drink would take that taste away, but replace it with something equally unpalatable. I raised the glass to my lips and started to drink.

As the countdown to Everest began, this was now how I was starting my days. The training routine I'd begun back in August was ratcheting up in intensity. I've always tried to keep fit and look after myself, as Lukas clocked when we first spoke, but in order to climb Everest, it was a different sort of fitness that I needed.

'You're not trying to get fit or cardiovascular fit,' Lukas had explained. 'You're trying to build your endurance.' As with his warnings about altitude, the mantra was the same: 'There are no shortcuts.'

Endurance training is about long efforts and low intensity; it's building up the body for a long haul, and there's no alternative to putting the hours in. On the mountain, Lukas explained, what kills you is your elevated heart rate. Because your blood becomes so thick, it tires your heart out. It just can't pump that fast. In the training, you need to learn to how to put your body through its paces without elevating your heartbeat.

The first step in that journey, literally, was to buy a stair climber for the office gym. It's a bit like a treadmill, but with steps that rotate round and round. You're continuously climbing, but going nowhere. For my training, it was invaluable, helping me learn to stress my body while keeping my heart rate low.

One of the other challenges with fitness for mountaineering is oxygen. Upper-body strength is important, for sure, but you don't want to bulk up. You need to develop muscle fibres that are tight, thin and long, rather than a load of muscle mass. Muscles consume oxygen, so the more muscles you have, the more oxygen you need. Body weight is important – the more weight you're carrying, the more energy you're going to need to get yourself up the mountain. But durability is more so, and that requires a different sort of training.

What you need to climb Everest is a combination of these different elements. You need the physical strength first. Then you need the cardiovascular strength on top of that, and finally the endurance level as well. That's not a conventional training set-up, but I was lucky to have a personal trainer, Ali, who devised and took me through a programme. It was hard work, and a lot of early starts – I'd be there in the gym at five or six in the morning, getting a session in before I started work.

The training was tough. There was a lot of step work. There was a high box that Ali liked to use, which he would have me stepping on and off. Another exercise I did a lot of involved a heavy, five-kilo ball, which I would throw again and again against the wall. All of these exercises were geared to getting my blood pumping and my heart rate up. But as the weeks turned to months, my heart rate began to slow down. Strenuous exercise, which sent the rate soaring at the start, I could now do with barely a blip.

I felt so much better generally. I was sleeping better, more deeply and felt more refreshed. Despite the early starts, I had more energy. My mental focus was sharper and I could see how my performance was improving at work. I began to understand how people became addicted to exercise: because you're feeling better, it pushes you to train even harder. My whole outlook on life had started to shift a little. Whereas it used to be Stephanie looking out for me at a dinner or a party, now it was me saying, come on, we should go.

My diet also changed. The olive oil and garlic starter was just the beginning of my new food plan. After my morning training session, I would have a bowl of oats with fresh almond milk. For lunch, it would be a salad or soup: the soup would never be packaged, but made from vegetables, slow cooked at a temperature under 42 degrees, so they maintained their raw nutritional value. For dinner, I ate what the family was eating, but it was nearly always vegetable dishes. I had plenty of supplements, too. I took magnesium, vitamin D and omega oils, all of which helped with my muscles, my cardiovascular system, and replenishing my minerals.

It was a lifestyle change, but because I could see the benefits, I didn't mind committing. Mountain or no mountain, the preparation by itself was changing my life for the better.

• • •

'OK, Hakan,' Daniel said, shaking hands with a glint in his eye. 'Let's see what you're made of.'

While Lukas was persuading Stephanie about Everest on the skiing trip to Lech, he wasn't leaving me idle. Instead, he had arranged for another of his guides to put me through my paces. So while Stephanie and Lukas enjoyed a good day's skiing, I was taken aside by the guide. Daniel was a mountain goat in human form. He'd been up Everest several times and, both literally and figuratively, knew the ropes. For all my training and scaling of Aconcagua, I still hadn't done that much in the way of actual climbing. Daniel wanted to see what I could do.

What that entailed was finding a 300-metre cliff and hanging me over the side of it, with just a carabiner and a rope for support. Carabiner is a German word, meaning spring hook, and is the small bit of kit at the heart of much mountaineering. It's what you use to attach yourself to the support rope to keep you from falling.

Finding myself over the side of the cliff, suspended by a rope, was a scary experience. Looking down beneath me, the ground seemed a long way off, and as deep as the snow was, it wasn't going to cushion my fall.

'OK, Hakan,' Daniel shouted down from above. 'Let's see how you get back up then.'

I looked at the wall of rock and ice in front of me. I didn't know where to start.

'Right,' I said. 'And how exactly do I do that?'

Daniel gave me a shrug. 'That's for you to work out.'

'Huh?' I felt a slight twang on the rope. 'But how am I supposed to get myself back up if you don't tell me how?'

Daniel gave me a smile and the same answer. 'That's for you to work out. Do you think anyone is going to tell you how on Everest? Well then.'

I knew I was being tested. It was a tough teaching technique, but I could see Daniel was smart and knew what he was doing. By stretching me, he was forcing me to learn, and learn fast. He wanted to test my temperament, too, to see if I'd panic when thrown in the deep end like that. But I took a deep breath and struggled on.

Lukas arranged for me to work with a number of experts. With Daniel, I did a lot of the more technical stuff, learning how to climb and descend with ropes: rappelling. I worked with an Italian mountain climber who was amazing at rock climbing, very strong and technical. And I went out with Alex, a member of Lukas's team, as well, a trip that I perhaps shouldn't have taken Stephanie on. We skied for several hours, then put the skis on our backs and climbed. It was both amazing and terrifying at the same time – we were climbing with no ropes, and so the exposure if we fell was huge. We ended up in places where no skier should ever go: 65-degree slopes on top of a south-facing, avalanche-prone mountain. But by the end of that day, and with the reports back from his guides, Lukas was becoming more and more convinced that I was putting in the time and work.

I had a head start in my learning from my experiences sailing. You can't be a sailor unless you know your knots, and so when it came to ropes, I knew how to tie and keep things secure. But that experience only took me so far. Looking back, I should have done more rappelling and on more difficult descents. I think my training didn't do that because the team didn't want to alarm me, but it meant that when I got to Everest, I found myself on open rock faces with a lot of exposure and not enough experience.

The northern route up Everest has three notorious challenges to overcome, known as the Three Steps. Out of these, the Second Step is the most daunting. Essentially, it's a cliff face, all but unclimbable, which is why most of the earlier ascents on the mountain took the southern route instead. But in the 1960s, the Chinese installed an aluminium ladder up the Second Step. This doesn't make ascending it easy, but it does make it possible. The danger from the Second Step is still there – the exposure from the potential drop, the risk from clipping on to the wrong line, one of those many old ropes there withering and unravelling. And if that wasn't terrifying enough, the Second Step had to be attempted at night in order for the timings to work. For all my bravado and excitement at climbing Everest, when I saw pictures of that step, my first thought was, there's no way I'm going to get up that.

In business, when I find myself facing a problem, I always tackle issues in a certain way: I identify the problem, break it down and familiarise myself with the challenges faced. I decided to approach the Second Step in the same manner. In order to overcome the problem, I needed to familiarise myself with it, and find a way of replicating the climb as best I could.

I called a Turkish mountaineer and explained what I wanted to do. He was happy to help, and together we hatched a plan. At the time, the new Istanbul Airport in Arnavutköy was in the final stages of its first phase of completion, as part of a long-term project to replace the Istanbul Atatürk Airport, which had long reached its capacity. The transfer of passenger flights to the new airport took place in April 2019, a month before my ascent of Everest, with the full airport completed in the mid 2020s.

Having such a large construction project meant there were also what are known as altitude technicians in the city – mountaineers, essentially, who were using their climbing skills to help in the construction of the airport roof. I hired those guys. I'm sorry to say I persuaded five of them to phone in sick one day and help me build a replica of the Second Step. About an hour's drive out of Istanbul, tucked away in the middle of an industrial area, is this remarkable canyon. It's well hidden: if you looked at the heavy industry surrounding it, you'd never know it was there. But it's a popular place with climbers in the know, with steep sharp sides to test your skills on.

Such was the height of the canyon and the size of the Second Step I was trying to recreate, we had to use a number of ladders lashed together to replicate it. It was a job and a half, and took them the best part of three days. Once it was complete, I looked up at it from the canyon floor below, a river (sadly polluted from the nearby industry) running behind me. My heart was in my mouth. It was a long way up.

Building the replica was only part of recreating the Everest experience. To do it properly, I had to not just get used to climbing it, but to do it in all my kit, exactly as I would on the day. I felt a bit stupid – not to mention extremely hot – putting on my summit suit, but it needed doing. I filled a backpack with the weight of the equipment I would be carrying. Most importantly, I needed to get used to scaling the ladder with my crampons on. Crampons are great for giving you grip in the snow; they're less useful when climbing a ladder, continually getting stuck and tripping you up.

The first time I tried to scale the ladder was a nightmare. It was scary as hell. I kept thinking to myself, am I really doing this? Trying to control those fears and concentrate at the same time was really difficult. I had to practise clipping in and out of the ropes as I made my way up. The crampons were awful: it was like being constantly tripped up, except rather than just falling over, the consequence was dropping all the way to the ground. More than once, I found myself scrabbling on the ladder for dear life as I lost my footing.

But the challenge with fear is to confront it. Having completed the climb once, and then collapsing as a nervous wreck, I had to brush myself off and do it all over again. It took a while, but each time I attempted the climb, it got a little easier to do. That's not quite the right description: the climb itself was as hard as before, but my confidence in my ability to cope grew. The fear was still there, and with every slip it lurched back into view. But I was better equipped to deal with it.

Having tackled the 'Step' in daylight, I then had to do it at night, as this was when I'd be scaling it on the actual climb. The one plus side was that it wasn't so hot now, wearing all my gear. But in every other way, the experience felt ramped up. In the canyon the lights of Istanbul felt a long way off; here it was properly dark, inky black. The only light was from my head torch, and a couple of climbers who had built the ladder watching from below. The beam of their spotlight helped, but only so much. I realised that sight was only going to be so helpful at night. I had to get used to climbing by feel, using my hands and feet as much as anything to guide me.

Halfway up the ladder, I heard a commotion below. More spotlights, more torches. I looked back down, but couldn't make out who they were. The shouts of my team were clear, however: I needed to descend. As I came back down, the new arrivals came into focus. These were members of the *gendarmerie*, the Turkish military police.

'Are you in charge here?' one of the officers asked me. 'What on earth do you think you're doing?'

'I'm preparing to climb Mount Everest,' I explained.

That answer didn't go down well.

'Really?' The military police didn't sound convinced. I think their suspicions were that we might have been preparing for something else, maybe something crime-related. Though it would be a strange criminal who wore crampons and a bright yellow summit suit. The more we talked, the more they realised I

Second Step training at Ballıkayalar.

was telling the truth, though they remained determined to maintain the upper hand.

'It's illegal to film here without a permit,' they told us. 'You're going to have to leave.'

I tried to protest, but they weren't the sort of people to argue with. I had to leave the canyon, my preparations incomplete.

• • •

'I've got a present for you.'

Having finished the day skiing in Austria with Lukas, he'd turned up back at the apartment where we were staying. As presents go, this was a big one: such was the size of the box that he could barely get it through the door.

'What is it?' my children asked, crowding around and curious.

'I'll show you,' Lukas smiled, and he dragged the box across into the living room. 'Is that a pull-out sofa?' he asked.

'I think so,' I said, confused. 'Do you want me to get it out?'

'Sure,' Lukas said.

'Are you planning on staying over?' I asked.

Lukas laughed. 'The sofa bed isn't for me, Hakan. It's for you.'

Opening the box, he took out and began to assemble what looked like a large transparent cube. It had a dark metal frame and was open at the base, with a plastic skirting to seal underneath.

'This is your father's hypoxic tent,' Lukas explained to the children. 'This is what he is going to be sleeping in over the next couple of months.'

It looked like some sort of mad scientist's contraption. 'You think I'm going to be sleeping in …,' Lukas gave me a look. Then I remembered what I'd felt like, that night on Aconcagua. 'This is what I'm going to be sleeping in,' I said.

'Come on,' Lukas gestured towards the sofa. 'Lie down. Let's see what this looks like for size.'

To the amusement of my children, I got a duvet and a pillow, and settled down as though I was going to go to sleep. Lukas lifted the tent over the top of me. On three sides it fitted well and sealed me in. The front was a bit more of a fiddle, which was where the plastic skirting came on, to form a seal over the top of the duvet.

'How's that?' said Lukas.

I gave him a thumbs-up and looked around. There was a spotlight nestled in one corner. Beyond I could see the curious faces of Lukas and my family.

'How often will he have to sleep in that?' Stephanie asked.

'Every night,' Lukas said.

Stephanie gave me a grin. 'I think you might be sleeping in the attic for the next few months.'

It took a while to get used to the hypoxic tent. I had a few things I'd take into it with me – water, ear plugs, a mask – but once you were in, you were in. And from the way it was set up, you couldn't really toss and move around like I was used to. It was possible to turn, but essentially, you had to learn to sleep on your back, which was something that I didn't normally do. For the first few weeks, I was exhausted through broken sleep. I'd get into a position, go down, then get uncomfortable and wake. It was a really stop-start experience.

Technically, it was tricky to get used to as well. The tent had an oxygen saturation meter, and I was constantly waking and checking the reading. Condensation was also an issue, and I was continually making adjustments so the outside world didn't disappear in a haze. And then, to add to it all, there were the flashbacks from Aconcagua. The higher the altitude setting, the lower the oxygen and the less you slept. I would wake up, gasping for air, and with a lurch would recognise those sensations. I had to work hard, mentally, to calm myself down from that and drop back off.

For the first couple of weeks, apart from the discomfort of trying to sleep in the tent, the effect seemed minimal. But as the tent began to replicate the atmosphere at higher and higher altitudes, my body struggled to adjust. Outside of the tent, by contrast, I felt amazing. It was almost like being superhuman: I could train better, and for longer. It got to the point where other people in the gym were stopping and coming over to watch me. I could go non-stop for an hour, my heart rate stable at 140.

One of the problems with the hypoxic tent was that you couldn't skip a day. A couple of days without sleeping in it, and you'd lose all your altitude training overnight. That was fine when I was at home, but such was my job that I needed to travel abroad for business. I went to one meeting in Cyprus and had to take the whole set-up with me. Put it this way, it doesn't exactly fit in your hand luggage. It took forever to get through customs, with the officers grilling me about what it was, and what certificates I had to use it.

I had another meeting in Japan I had to go to. Rather than going through all that again, I found a tent that I could hire. The hotel I was staying in were able to help and, when I got there, the rented tent was set up. Rather than being a small box, this tent was huge, covering the whole bed in a sealed cylindrical chamber. Great, I thought. I had a really important meeting in the morning, and for the first time in months I could get a decent night's sleep. I could stretch out, rather than being stuck on my back. I put my ear plugs in and eye mask on and looked forward to getting some decent rest.

In the middle of the night, I came to, completely disorientated. I was sodden and gasping for breath. I was struggling to work out where I was. I felt faint. I ripped off my eye mask and blinked in the glare. The lights on the tent were flashing red. I took out my ear plugs and could hear an alarm going off, an automated voice shouting what I presumed was a warning message in Japanese. Beyond, the room was pitch black. My chest was tight, and I realised I was suffocating. The oxygen saturation monitor was flashing with an error message and didn't respond when I tried to adjust it.

I had to get out, but I couldn't remember how. The whole point of the tents is that they need to be tightly sealed to work. My hands were feeling along the side to find a catch, a handle to pull to let me out. All the time, my lungs were winding up tighter and tighter. I was starting to feel dizzy and light-headed. Somehow, and I'm still not sure how, I found my way out. As the seal broke, my lungs sucked in huge breaths of hotel-room air. I half crawled out of the tent, like someone out of a car wreckage, and lay there, sodden in sweat.

Later, I learned that there had been a power surge. When there is a power cut, the hotel generators would kick in in response. But because it was a surge rather than a cut, it knocked the generators out. Once that happened, the tent stopped pumping oxygen in, and there was only so long that I could survive. And although the warning system on the tent kicked into action, my ear plugs and eye mask blocked them out. I'd had a lucky escape – another few minutes asleep and I'd have been unconscious.

• • •

As the weeks and then months ticked down, my training began to come together. Physically, I was a different person to the one on that Greek island the previous summer. I could feel the strength I'd gained in my upper body, but also the leanness, rather than the muscle I'd built up. I'd been up my ladders in the Turkish canyon enough times that I was no longer dreading the thought of the Second Step, and in fact was beginning to look forward to the challenge. And as long as the electricity lasted the night, the hypoxic tent was clearly having an effect as well. I could sleep in air mixed to simulate what it was like thousands of metres up, and wake up with no side effects. Despite of all this, I still had my doubts. The way my knee had gone on Aconcagua had worried me. When I'd arrived back in Turkey, I'd had it checked out. The doctor had taken an X-ray and told me that I had a torn meniscus. We discussed whether an operation was a possibility but instead decided on a plan of building up my strength and training my body to resolve the issue. As I started the gym sessions and preparations in earnest, I was initially worried my knee might go again. But it was OK. The longer the training went on, the more confident I got. But even so, the nag in the back of my mind refused to go away completely. What if it went on Everest?

That fear wasn't helped by the story of Beck Weathers. Beck Weathers, as I mentioned earlier, is one of those Everest tales that even those who know little about mountaineering know about. He wrote a bestselling book, *Left for Dead*, about his experience. Beck had scaled Everest in 1996, when the group he was in was struck by a blizzard. When the storm lifted, Beck was in such a weakened condition that he was left while others went for help. When help arrived, they decided that he was too close to death to be able to make it off the mountain. Somehow, he survived, and made it down. But his survival came at a cost: part of his right arm and feet, his fingers and thumbs on his left hand were amputated. His nose was reconstructed using tissue from his ears and forehead.

The guides in Aconcagua knew Beck, and through them I invited him to come to Istanbul and speak at a Young Presidents' Organization (YPO) event I was involved in. For some reason, and I'm still not quite sure why we thought this was a good idea, the talk took place in the cool room of a chocolate factory. It was freezing in there, and everyone had to wear these big parkas to hear him speak.

Beck was an extraordinary speaker. Struggling to hold back tears, he described how when he'd been left for dead, stuck out on the mountainside at minus fifty, the only way he'd been able to keep going was to think about his family. How he got down was little short of a miracle. He'd been blinded by a combination of high altitude and exposure to ultraviolet radiation: the only way he was able to walk down was to follow voices that were carried by the wind. It sounded incredible when he told it, and having been to Everest and seen the terrain for myself, it seems all the more extraordinary that he didn't die.

It was an emotional speech. Beck kept on returning to the same point, that he'd learned from the experience about how the value in life was all in your family. What I clocked in that cool room was that the person who is not supposed to be on the mountain is going to die on the mountain. If you're not ready, if you haven't done the prep, if you have any physical issues, then you shouldn't even consider going on the climb.

Beck's talk took me back to another conversation I'd had on Aconcagua with one of the guides. After we'd summited and made our way back down to camp, I was in the full flush of success. I told him of my plan to climb Everest the following spring. The guide, who'd been up Everest himself, looked as though I was mad.

'No,' he said simply. 'No way. Aconcagua is difficult, and you've had a tough experience here, but Everest is something else. Seriously, Hakan, you have no idea. Go off and climb ten other mountains first before you even think of doing that. If you try and climb Everest, you'll die.'

At the time I remember being quite shocked at the guide's vehemence that I shouldn't go. When I spoke to Lukas about it, he was more reassuring. 'A lot of guides think like that,' he said. 'They don't understand the support we give you, the infrastructure we have in place. Also that we're going to train all your weaknesses out of you before you set foot on the mountain.'

At the time, Lukas's words reassured me. But as the deadline for departure ticked close, went from months to weeks to days away, it was the words of the Aconcagua guide that came back to me in the quiet of night. *If you try and climb Everest, you'll die.* As much as I'd put the hours in to train, those words continued to haunt me. I believed in what Lukas told me, but what if he was wrong and the Aconcagua guide was right?

A TIDAL WAVE OF PLASTIC

One Christmas and New Year, I took my family to Thailand on holiday. Stephanie is Swedish, and from a culture of going somewhere warm in the winter. We'd just had another baby and it seemed a good idea to enjoy some winter sun to recover. Thailand is a place that I can take or leave. It's one of those areas that over the years have become increasingly developed and full of tourists (I appreciate that I'm saying that while I was going there and adding to those numbers!). It's not a place where you can go and teach your children about the world – I'm not a fan of how they look after animals there, and didn't want to show my kids tigers in cages or elephants being mistreated.

What I was more excited about was Maya Bay. When I was living in Hong Kong this was somewhere that I'd visit on my sailing weekends. I remembered well the crystal clear nature of the water, the creamy whites of the beaches, the rustle of palm leaves. Maya Bay is one of those places where it feels as though you leave the world behind, step away from the tourist trail, re-immerse yourself in nature and find yourself again.

I was looking forward to this part of the trip. Beforehand, I'd talked it up to my children, shown them photos of my previous sailing trips. I told them a story about an island in a beautiful bay where pirates used to hide, and if we went to search the caves, maybe we'd find some buried treasure. I had a plan to distract them and head into the caves ahead alone to hide something for them to find. It was all set up: one of those shared family memories that we'd remember for the rest of our lives.

We would certainly have an experience none of us would ever forget. Just not in the way I'd hoped for.

Even before we reached the bay, I could see that things had changed from the last time I'd sailed there. We'd gone early in the morning, at a time before the normal tourist boats were setting off. But as we made our way across the Andaman Sea, the waters were already clogged with other vessels. It seemed that since I'd been away, others had found my idyllic hideaway and discovered it for themselves.

We made it into the bay and I dropped anchor just short of the shore. I knew this was a place where there were no sharks or jellyfish, and my children could enjoy the experience of swimming to land. But as we made our way through, the sea was far from crystal clear. You know that scene in Star Wars where Luke Skywalker and Han Solo find themselves in the garbage compactor? It was a bit like that, wading our way through all the rubbish.

The shoreline, rather than being pristine white, was knee-deep in detritus. Plastic. Styrofoam. Lighters, water bottles, hairbrushes, clothes pegs. Broken-down boat parts and fish-box parts. Plastic forks. Straws. Bags. I didn't have anything on my feet so had no choice but to step my way through. I picked up my two children, one on my side, the other on my shoulders, and carried them through.

It was disgusting. Some of it was still recognisable, the waste in its original form. Other parts of it had decomposed in the sun and mixed with the sand to make a revolting stew. I wasn't sure where to put my feet. I didn't want to look down at what I was standing on, but at the same time, I was worried I might step on a nail or something sharp. As I looked down, I could see that the waste was teeming with insects, centipedes and other creatures crawling over my feet. There were birds, too, the carcasses of gulls and other species lying motionless in the mess.

'Baba,' my daughter asked me, 'what is all this trash?'

There's a famous moment at the end of the film *Planet of the Apes* when the protagonist, George Taylor, played by Charlton Heston, discovers the remains of the Statue of Liberty on a beach. Taylor sinks to his knees in the realisation that rather than being on another planet, he has been on earth all along, and that mankind is responsible for the destruction of the world as he once knew it. 'You maniacs!' Taylor shouts, slamming his fists into the sand in despair. 'Damn you! God damn you all to hell!'

At that moment, the Statue of Liberty seemed about the only thing that hadn't been washed up on the beach in Maya

Bay. But I completely understood the despair, hopelessness and anger that Charlton Heston's character felt. It was shocking to see somewhere so beautiful so ravaged, and for humankind to be so blatantly responsible.

'Baba,' my daughter asked again. 'Why is all this trash here?'

I shook my head, tears in my eyes. 'I don't know,' I replied softly.

• • •

In another late-1960s Hollywood film, Dustin Hoffman famously plays the role of Benjamin Braddock, the graduate of the film's title, freshly home from college and both uncertain and ready to step out into the world. At a party hosted by his parents to celebrate his graduation, Braddock is pulled aside by a family friend, Mr McGuire, to be offered a piece of friendly advice.

'I want to say one word to you, Benjamin, just one word,' McGuire says. 'Plastics.'

Had Benjamin Braddock followed McGuire's advice and gone into plastics, he might well have made his fortune. Back in the 1960s and 1970s, plastic felt like the material of the future. Strong, pliable and long-lasting, it seemed a technological solution to many of the world's needs. At the time that McGuire was offering his advice, plastic was still a growing phenomenon: the world was producing about 25 million tonnes of it a year. It wasn't until the middle of the 1970s that production reached 50 million tonnes – a level that still allowed the world to deal with the waste relatively straightforwardly.

What has happened since then, however, sees the graph of global plastic production shoot up in the same way as the earlier ones mentioned about population and energy growth. By the 1990s, both plastic production and plastic waste generation had tripled in two decades. In the early 2000s, our output of plastic waste rose more in a single decade than it had in the previous forty years. In the last fifteen years, we have made as much plastic as all our previous production added together. In total, researchers estimate that we have made more than 8.3 billion tonnes of plastic. As Lucy Siegle puts it in her book, *Turning the Tide on Plastic*, that's the equivalent of 1 billion elephants. To continue the weight analogy, we're now producing over 300 million tonnes of plastic every year. That's nearly equivalent to the weight of the entire human population.

It's not just the volume of plastic produced that is unsustainable, it's also the way we deal with the waste. By 2015, Siegle notes

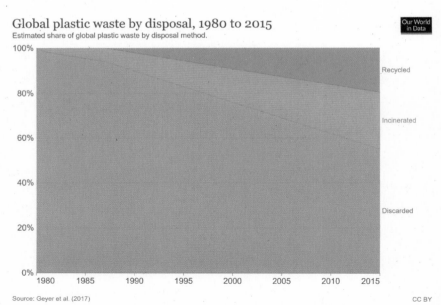

Global plastic waste by disposal, 1980 to 2015
Estimated share of global plastic waste by disposal method.

Our World in Data

Recycled

Incinerated

Discarded

Source: Geyer et al. (2017)

CC BY

A very limited portion of plastic waste goes to recycling.

that just 9 per cent of our 8.3 billion tonnes has been recycled, 12 per cent has been incinerated and 79 per cent has been accumulated in landfills or the wider environment. The vast majority of the plastic produced, which we're adding to every year, is still here. At the moment, the estimate is that 8 million tonnes of plastic end up in the oceans and waterways every year. As Siegle puts it, 'That is the equivalent of a truckload of plastic being upturned and shaken out straight into the sea every minute of every day … By 2050 the ocean will contain more plastic by weight than fish.'

The plastic in the sea tends to have been washed into the oceans via rivers. About half of the plastic found there is thought to come from just five countries: China, Indonesia, the Philippines, Thailand and Vietnam. According to research for The Ocean Cleanup, just 1,000 rivers account for 80 per cent of the plastic waste, with the remaining 20 per cent coming from an additional 30,000 rivers. The largest polluting river is the Yangtze, from which an estimated 330,000 tonnes of plastic waste ends up in the sea each year. But this is far from just a Chinese issue: the Ganges, Cross, Amazon, Brantas, Pasig, Irrawaddy and Solo are also in the top ten.

So, while some of the worst cases are in Asia, plastic is still heading into the sea from wherever you might be reading this. And even if you're one of the good people who are consciously recycling their plastic, remember that much of that ends up elsewhere to be disposed of. A bottle recycled in Europe, say, might still find its way into the Pacific, depending on how that waste is being dealt with.

In 1997, the oceanographer and yachtsman Charles Moore was sailing back across the Pacific to the United States after taking part in that year's Transpacific Yacht Race. As he was passing the North Pacific Gyre – a large, stationary patch of water surrounded by a circle of currents and winds – he became aware of the rubbish he was sailing through. 'I was confronted, as far as the eye could see, with the sight of plastic', he wrote for *Natural History Magazine*. 'In the week it took to cross the subtropical high, no matter what time of day I looked, plastic debris was floating everywhere: bottles, bottle caps, wrappers, fragments.' The phenomenon became known as the Great Pacific Garbage Patch.

The rubbish that Moore witnessed had made its way there from both the east coast of Asia and the west coast of North America, together with further debris that had been left behind by ships. It is estimated that, depending on currents, it takes six years for the rubbish to make its way there from America, and a single year from Asia. The patch covers a size of 1.6 million square kilometres, about three times the size of France. Recent estimates suggest that it contains 1.8 trillion pieces of plastic, about 80,000 tonnes in weight (the equivalent of 500 jumbo jets, if you'd like another analogy). This number – which equates to 250 pieces for every human on the planet – is described as a mid-range estimate. The actually total could be as high as 3.6 trillion pieces of plastic.

And much of this is only the plastic that we can actually see. More difficult to deal with still is the emergence of microplastics. Microplastics are pieces of plastic that measure less than 5 mm. Some microplastics are formed by larger pieces of plastic breaking up. But others are made intentionally small: microbeads, which can be found in facial scrubs, for example, or industrial abrasives used in sandblasting. Then there are microfibres from clothing, and nanoplastics too – these are formed from microplastics breaking up. They're so small, essentially like dust, that they're all but impossible to keep track of and separate out. It's one thing to take a bottle out of the sea – but once it has broken down into nanoplastics, it's mixed in with the water for good.

Over the last few years, scientists have found evidence of microplastics in the most surprising and isolated places on earth. In 2018, microplastics were found in the Mariana Trench in the western Pacific – at almost 11,000 metres, the deepest point of the ocean. At the other end of the scale, in November 2020, researchers revealed that microplastics had been found on Mount Everest, just short of the summit at 8,440 metres high. Lead researcher Imogen Napper took snow samples at eleven separate points up the mountain and was 'really surprised … to find microplastics in every single snow sample I analysed … To know we are polluting near the top of the tallest mountain is a real eye-opener.'

Six months earlier, microplastics were found in sea ice in Antarctica. The fact that the ice analysed was drilled in 2009 clearly suggests that it has been in evidence there for a while. This mirrors the discovery in 2018 of microplastics in sea ice in the Arctic. Perhaps most disturbingly was the revelation in December 2020 that microplastics had been found in the placentas of unborn babies. Analysis of the plastic particles suggested they had come from packing, cosmetics and personal-care products.

Such was the size of the plastic particles – 0.01 mm – that they are small enough to pass into the bloodstream. 'It is like having a cyborg baby,' Antonio Ragusa, who led the study, was quoted as saying: 'No longer composed of only human cells, but a mixture of biological and inorganic entities.'

The effect of all this requires further research, as do the effects of microplastics we all now regularly consume. We start young: an October 2020 study showed that bottle-fed babies are swallowing millions of microplastic particles every day. Plastic teabags have been shown to release 11.6 billion microplastics and 3.1 billion nanoplastics into a single cup of tea. Even if you don't drink tea or baby milk, research in 2019 showed that humans consume at least 50,000 microplastic particles a year after an analysis of human faeces. Tap water, seafood, beer, salt and sugar were just some of culprits.

The health effects of all this are still being researched.

• • •

'There is a scene in *A Plastic Ocean* where I open the stomach of a shearwater bird and all this plastic comes out of its stomach. That, without a doubt for me, was the critical part of the film for me as the storyteller, but also for the audience.'

When the documentary film *A Plastic Ocean* was released in 2016, it was hailed by Sir David Attenborough as 'the most important film of our time.' Seven years in the making, the film took viewers around the globe, looking at the effects that plastic was having on the natural world. Craig Leeson was the film's director. A long-standing journalist and documentary-maker, even with all his experience he was still shocked at what he saw. In one of the key scenes, the film follows a seabird biologist on Lord Howe Island in the Tasman Sea, who collected dozens of dead shearwaters every day. In the stomach of one particular bird, Craig counted 234 separate pieces of plastic.

'I knew the statistics of how 96 per cent of seabirds were ingesting plastic,' Craig told me. 'But when you see it yourself for the first time, it's shocking because you see bits of plastic coming out. I saw a red bottle cap that came from a Coca-Cola bottle. That could have been from a Coke bottle that I drank from when I was drinking Coke and using plastic bottles. That was five years earlier, but it could have been one that I had used and thrown into the environment that the bird had eaten. If it wasn't that particular bird, chances are that my bottle top was in a bird or an animal somewhere on the planet, and it had caused that creature either incredible pain, or had choked it to death. That was the moment,' Craig remembers, 'that I felt personally responsible for the problem.'

The effect on wildlife of the plastic problem is scandalous. It is estimated that over 100,000 marine animals are killed by plastic each year. And out of the species affected by ingestion and entanglement from marine litter, 15 per cent are endangered. If you wonder why animals eat such items of plastic, it's because having spent years in the ocean and becoming covered with algae and fish, the plastic smells to them like actual food. No sea creature is too large to die from plastic pollution: in 2018, the body of a sperm whale was found washed up on the beach in southern

Spain. An autopsy found 29 kg of plastic in its stomach, including plastic bags, nets, ropes and even a jerry can.

As well as Craig, I also discussed the film with the husband and wife team of Julie Andersen, the CEO of Plastic Oceans International, the not-for-profit organisation behind the documentary film, and William Pfeiffer, the Executive Chairman of Globalgate, who was involved in the production of the film. 'What worries me about plastic is the global amount,' Julie told me. 'The unlimited production of plastic that continues to grow without any waste management backstop. What worries me is that because it's so big, we're not participating in the right conversation. It's like, "let's keep waiting for the innovation that is going to solve the world's problems." This approach belittles our individual capacity to innovate. We don't need a $1 billion R&D investment to create something that is $100. It is about empowering these individual communities that truly will innovate and stop the flood of plastic.'

William cited the example of the success of their son in a local campaign. 'Our son went to social-justice camp in Santa Monica where they had to pick one project to focus on to get active in the community during the summer. Ethan was nine years old and came up with the idea of banning single-use plastics. The children marched to the city hall, gave speeches, and sang songs. The government officials came out, and they put the proposal on the agenda for the next meeting, and the following month, they banned plastic straws in Santa Monica.'

'Creating this ban on a local level was much more effective,' William explained. 'Going straight to the capital of California and trying to get them to pass the same ban on the state level, you're going to face big lobbyists from Dow and from the oil companies.' He explained how Plastic Oceans were involved in a project on Easter Island, that, despite its remote location, struggles to cope with plastic pollution. With a growing population and a huge increase in tourist numbers, not to mention what is being washed up on the shore, the island suffers from one of the highest concentrations of microplastics in the world.

'On Easter Island, we work with a local music school that has built the school out of recycled materials: glass bottles, tyres and other waste. And as well as teaching music, they teach the children how they built the school. They also grow their own fruit and vegetables and teach the children about that. Even though their focus is on music, they incorporated different elements of sustainability within that practice.' As William sees it, 'we are one global society with many local communities. I hate to use the term brush fires, but brush fires are what we're trying to create, to build the real bigger fire of change.'

It's an interesting philosophy and one that feels both engaging and empowering: that while the overall problems might feel insurmountable, on the ground we all have the ability to make a difference.

A Plastic Ocean is a film that has lit brush fires large and small (if you haven't seen it yet, you can find it on Netflix, among other distributors). Craig gave me one final great example of how a project like this can make a difference on a personal level: 'I moderated a panel discussion at a sustainability forum in Singapore last year. On that panel I have the VPs of Danone, Exxon and all these companies that produce a lot of the products that are a problem. One of the VPs came up to me afterwards and said, "My kids watched your film. They came home to me and said, 'Dad, why do you work for a company that's destroying the environment?" And he said, "I had no answer." He said to his children, "What are you talking about?" And they made me watch the film.'

'The VP said, "I went to work the next day. I sat at my desk, and I just sat there for half an hour staring at my desk just wondering what the hell I could do to turn my kids' thoughts about me around." Because every parent wants to be a hero to their kid, right? They don't want to be a villain. That's where storytelling becomes so powerful: you can tell a story that kids grasp hold of. They go ahead and question their parents, and their parents then come and question me about what they could do. And that's a very powerful thing.'

. . .

The first type of plastic invented was a substance called Parkesine, back in the 1850s. Its creator was the British inventor Alexander Parkes, who made his substance from a mixture of chloroform and castor oil. A guide to the 1862 London International Exhibition, in which he won bronze for his discovery, described his substance as 'hard as horn but as flexible as leather, capable of being cast or stamped, painted, dyed or carved'.

One of Parkesine's first uses was in clothing. The fashion for Victorian shirts required the use of materials such as cow horn, ivory and tortoiseshell – these helped satisfy the vogue for stiff collars, while the desire for tortoiseshell buttons saw the mass culling of hawksbill turtles to sate demand. In an ironic twist on how the relationship between plastic and nature would develop, by using Parkesine in shirt manufacture, the plight of the hawksbill turtle was, momentarily at least, lessened (sadly, that reprieve hasn't survived, and the hawksbill is currently critically endangered).

The relationship between fashion and plastic has continued, but now, rather than helping the environment, it is serving to seriously threaten it. This is because of the use of man-made fibres in many items of clothing. One of the biggest culprits is polyester, which can be found in about 60 per cent of modern clothes. In some respects, these synthetic materials have environmental advantages: compared to manufacturing cotton, their production requires less water and pesticide use. But on the (very) down side, their manufacture produces three times as much CO_2, with the kicker of the non-biodegradable waste that is left behind.

Every time you wash an item of clothing made of synthetic materials, it releases microfibres into the water. These are tiny strands of plastic that gets washed down the drain and into the rivers and oceans. To complete the journey, they are digested by fish, travel back up the food chain and, yes, are eventually consumed by us. In 2016, Plymouth University undertook a twelve-month project washing different kinds of synthetic materials at different washing machine temperatures. They discovered that each cycle of washing could release up to 700,000 synthetic particles per wash. Acrylic was the worst culprit, followed by polyester-cotton and polyester. Their findings echoed another by the University of

California, Santa Barbara, which found that washing a synthetic fleece jacket released a staggering 100,000 microfibres on average. A 2020 study by the same university found that over the previous twelve months, the amount of microfibres released into the Californian environment alone was a massive 13.3 quadrillion. If that's a hard number to wrap your head around, the report's authors pointed out that this is 130,000 times the number of stars in the Milky Way.

The challenge here is what we call fast fashion – clothes that are cheaply made, worn for one season and then discarded in favour of the next look. To compete in this industry, costs have to be low. For the manufacturers, there is little incentive to make more durable and long-lasting clothes – they are more expensive to produce and would result in consumers needing to buy fewer items. For the fashion-conscious consumer, it also creates difficult choices.

As one of the leading global manufacturers, I wanted to see if there was something we could do to stop these microfibres getting into the water. Getting manufacturers and consumers to change their habits will take the sort of time we don't have, but was there something that we could add to our washing machines? At Arçelik, we have an amazing research and development team, so I asked them: is there a way our washing machines can tackle the issue? Our team did an amazing job, and came up with the idea of a filter that removed the microfibres. It took a lot of time to get this to a level ready to go to market – it took us two years, in fact. We had to test the filter rigorously through a number of cycles, to see how long it lasted before it needed to be replaced. We also had to redesign the washing machine to accommodate the new filter, creating a new pump and cycle system. All the algorithms in the machine had to be rewritten.

Once we knew how it worked, we then had to see if we could bring the cost down to make it viable. Our market is very cost-competitive, so passing large costs on to the consumer will depress sales and reduce the use of the new technology. The early filters cost €25 to produce, but with hard work, we were able to get the filter cost down to €2 or €3 per unit. At that level, the retail price of the washing machine is pretty much unaffected. There will still be an onus on the consumer and the retailer to play their part. Once the filter is full, they will need to be replaced. And the filter needs to be properly recycled: if they're just thrown away, they will end up in landfill. And replacement filters will need to be readily and easily available. These washing machines are still relatively new on the market, so we'll see how it works in the next couple of years.

We brought in one other element of innovation into this process. Usually, on a project like this, you want to enjoy the returns for the R&D you've put in: a new product gives you that market edge over your rivals. But in the case of the washing machine filter, we decided to make the intellectual property open-source. When I first suggested it, some people in the company thought I was crazy, that we'd be throwing away our hard-won advantage. But it felt the right thing to do for the environment, and who knows? It may encourage our rivals to follow suit.

As a result, we've announced that we are ready to share our technology with all our competitors. Others are watching and waiting to see what happens. But if the microfibres filter is a success, then I'm sure they will come on board as well. For me, as a business leader, it's a trade-off – what we lose in market advantage, we gain in brand awareness. And the big winner in all of this is the environment. And while the washing machine filter will only dent the plastics problem in one particular way, it's an example of how we can all work together to play our part.

And when my children ask me what I'm doing to stop the flood of waste we waded through in Maya, I know that I'll be able to look them in the eye.

NEPAL, TIBET AND TENSION

Santa Maria del Camí is situated in the centre of Mallorca. It's a beautiful Spanish village, its buildings that typical mix of whites and soft oranges, stone and exposed brickwork. Above, the clouds bustled against the brilliant Balearic blue of the expanse of sky. I adjusted my bicycle helmet and pushed off for the mountain roads beyond.

Away from its beaches, Mallorca is a cyclist's dream, its inland roads comparatively free of traffic, offering both a testing workout and remarkable views. As I cycled away from the village, the road took me up through the local vineyards and into the hills. Six months earlier, I might have felt the pull of the slope as I cycled up, the thinning of the air in my lungs. But after all my training, I barely noticed either.

It was my last few days before setting off for Everest. My gear had been sent on in advance to Nepal. I was spending a last few days of holiday with my family before setting off. The training was done now, but I couldn't stop myself from still trying to squeeze in a bit more exercise. My levels were all OK, with one exception: I was still scoring low on the endurance targets. My trainer had taken me to the Marmara University athletics track, and had timed me against Lukas's criteria. Each time, I came up short. I knew I had the power I needed, but the consistency was harder to acquire. My friends Fred and Marco, who I was going to meet in Dubai on the way, were both marathon runners, so for them it wasn't an issue. In a way, my skill set for Everest was the opposite: I'd done more training and had more upper-body strength. But they had that experience of years of long-distance running. That, I couldn't compete with.

Ahead of me on the road was a group of female cyclists. You could tell from the way they were kitted out that they were serious about their riding. Good bikes, lycra, aerodynamic helmets. The way that they were tucked in behind each other on the road, each taking turns at the front then dropping back, showed that they meant business. My lack of proper kit and ordinary bike looked second-rate by comparison. But to my surprise, I was gaining on them. And even though the road was ratcheting up in gradient, the gap continued to close.

If I hadn't done my training, I knew that they'd be leaving me for dust. But I was in good nick. I nodded and said *hola* as I overtook them, pulling clear well before we reached summit. Below, Mallorca spread out before me. Over the top, I started to freewheel back down towards the villa where we were staying. My father-in-law was with us, and he was an amazing cook. After all the careful dieting, I was allowing myself to eat well for these last few days. A couple of extra kilograms, I knew, would be a good thing – fuel for the journey ahead.

If only Everest was as easy as the mountain I'd just ridden up, I thought.

• • •

From Mallorca I flew to Barcelona, and from there on to Dubai. It was here that I was meeting Fred and Marco and together we'd fly on to Kathmandu, to meet up with the rest of the expedition team.

Fred and Marco are two great friends who I have known for years. It is one of those friendships where we are all quite different people but somehow get on really well. Fred I first met through another friend, Sebastian, who is one of my oldest friends. I met him when I was twenty-one in Hong Kong. When he moved back to Europe, he introduced me to his circle. We did a lot of stuff together as a group, including going on summer and skiing holidays. Sebastian is a close friend of Fred's, so that's how we met.

Fred is an entrepreneur, and an incredibly successful one at that. He was one of the first people in Europe to notice how US firms had started outsourcing their call centres. When he imported that process over here, he was ahead of the game. Through a mixture of management buyouts and low-cost funding available at the time, he acquired a number of businesses to become the biggest call centre provider in the world. He's very skilled at buying businesses and re-leveraging.

Fred likes his projects. He's now building the world's largest catamaran, at sixty metres in length. The design is amazing; I want to call it a superyacht, but he prefers to describe it as a museum.

He wants to use it to bring art to a wide audience, sailing exhibitions from port to port. He's a big traveller as well as a sailor. As I'm writing this, he is in a hospital in Italy, recovering from severe trauma to his lungs and fractured ribs. He was racing a foiling sailing boat and crashed. He flew across the boat from the helm and slammed into the winch. He was barely able to speak on the phone, experiencing the worst pain in his life. Yet I'm sure he'll be back at it again – soon. He is also a helicopter pilot. And as well as being a helicopter pilot, he races cars. He likes to see himself as a bit of a modern-day Rubi Rubirosa.

He's very driven and focused on himself. On the mountain, he would go at his own pace, whatever anyone else was doing. He'd use as much as oxygen as he wanted. He liked the fact that we were there, but in strange way, he was actually there alone. Seven years before, he'd tried to climb Shishapangma, which is another 8,000-metre peak. He got so close to the top – within a couple of hundred metres – before they had to turn back. So on this trip, he was determined to make it, whatever it took.

Marco's surname is de Longevialle, meaning, in old French, the long valley. He co-founded a company which Fred and Sebastian put money in. That was based in Paris. He became CFO and then deputy CEO. He's a bit of an old-fashioned ladies' man and is always very conscious of his appearance. He always has a handkerchief in his pocket and is never anywhere without a jacket. Even at Base Camp, he had a jacket with him! He's perhaps the nicest travel companion that anyone could ask for; he's very gentle, very polite, very considerate.

Fred and Marco are both great fun in their different ways. Together, they're a bit like an old married couple, always arguing over this and that, except in a playful way. By the time I found them at Dubai Airport, I could hear them having one of these conversations even before I saw them. We all hugged, and I could feel that ripple of excitement immediately. We were actually going to do this.

'So how have you been?' I asked.

'Tell him about Rio de Janeiro,' Fred nodded to Marco.

'That was amazing.' Marco was looking more tanned than usual. 'It was a friend's bachelor party. A fortnight ago, but I can still feel the effects.'

Marco proceeded to tell me about this crazy night. At each mention of smoking, drinking and womanising, my eyebrows went up further. I thought of my own training, and how abstemious I'd been. And here was Marco, who seemed to have included partying in his training routine.

'Relax, Hakan,' he slapped me on the back. 'It'll be fine.'

Marco had that breezy confidence to him, that life would somehow turn out OK. I hoped he was right. I knew he was fitter than me, as Fred was, so maybe he could get away with his partying. But I put my worries aside as the three of us chatted, waiting for our connecting flight. With the jokes and boys' talk flowing, it was almost like just another trip. Almost.

• • •

Although we were taking the north route up Everest and setting off from Tibet, we still had to go via Nepal. In Kathmandu, there was a visa and insurance to sort out, together with the climbing permit and official clearance to travel to Tibet. They do a security check: given the political sensitivities in Tibet, the Chinese are strict on letting anyone in they believe to be supportive of Tibetan separatism.

For me, I had the added complication of being Turkish. The Chinese treatment of the Uighurs is well-documented and, given the close ethnic and cultural ties, their plight is a source of ongoing tension between China and Turkey. I was lucky that having been born in Norway, I've always been in possession of two passports, and it was on my Norwegian one that I applied for my permit. Even so, I'd been to China on a number of occasions for work with my Turkish passport, and knew those visas would come up on the embassy computer. Not only were there Chinese visas, but the computer would also flag up the fact that I'd once been banned from entering the country. This was due to having taken part in a trip to Xinjiang, when President Erdoğan led a delegation of around 300 Turkish business leaders to improve trade with the region. Following that, tensions rose between the two countries and for about a year and a half, I was unable to get a visa to visit mainland China. Relations had improved since then, but I knew it would only take one hostile member of embassy staff in Kathmandu and my trip would be cancelled.

When Fred, Marco and I arrived in the Nepalese capital, we met up with all the members of the group expedition. There were a number of familiar faces. Firstly, there was Lukas, looking fresh, fit and relaxed. There was Leo, our high-altitude doctor, who specialises in oxygen and tissue saturation. When he's not doing that, he's something of a brain surgeon, highly skilled in complex surgery. I was hoping he wasn't going to need those skills on this trip.

We also had a Chinese businessman from Phoenix who was going to take part in the first part of the expedition: he wanted to come to Base Camp and then head back again. He'd had a battle with cancer which he'd overcome, and this was on his bucket list. There was a Swiss businessman, who was one of the country's biggest food importers. He loved the mountains but had lost a couple of people close to him through climbing accidents. So his journey had a powerful, personal element to it.

Finally, there was another friend of Fred's, Marco's and mine. Gabriel Picard is in the wine trade and is one of the leading wine producers in the world, producing millions of bottles of wine in France each year. They've expanded the business out into the US, where they've gone into wholesaling and liquor. Gabriel is a lovely guy, a very dry Frenchman with a heart of gold. Throughout the journey I felt he was always looking out for me, ready to help if I needed it.

We stayed in a hotel in Kathmandu that was crawling with climbers. There were groups like us who were waiting for their visas and travelling to Tibet. But the majority of the people staying there were planning on climbing Everest on the Nepalese side. The southern route is the way that Sherpa Tenzing Norgay and Edmund Hillary scaled the mountain, becoming the first people to reach the summit on 29 May 1953. It's the more popular route: while the Chinese issue just 300 permits each year for those wanting to climb Everest, on the Nepalese side, there are almost four times as many available. Looking around the hotel in Kathmandu, it felt as though it was rammed with mountaineers. Everywhere you saw these little huddles of people, all wearing the same gear and all of a similar build. It was as though there wasn't an ounce of fat to be seen. Everyone had that sinewy leanness to them.

While we were waiting for the permits to come through, we visited the mountaineering stores. These were as packed as the hotel. Although I had all my gear with me, Fred and Marco still had plenty of stuff to buy. So I went with them, and inevitably ended up buying more equipment. I think that was a nervous reaction, as much as anything, to reassure myself that I had all the kit I needed.

The following morning, we headed to the airport. The airport was crammed; it was noisy, dirty and almost dangerously over-crowded. The flights out were all overbooked, making it a bit of a bunfight as to who was getting on and who was getting kicked off. Unlike on my flights across from Spain to Nepal, I now had all my gear with me. We had this fleet of trolleys, all loaded up with a pile of massive duffle bags to try and navigate the crowds with.

We checked in and sat down to wait for our flight. There was no catering, no coffee or food to buy. Slowly, the hours ticked by. Fred, Marco and I would usually be quite chatty, but our conversation dried up and we idled away the time playing backgammon on our iPads. After six hours of this, Lukas came over to us with a grim look on his face.

'There's bad weather,' he said. 'The flight's been cancelled. I'm afraid we've got to go back to Kathmandu and try again tomorrow.'

That was a deflating moment, and worse was to come: when we got back to the hotel, our rooms had gone to another set of climbers and we had to scrabble around for alternative accommodation. That evening at the group dinner, someone ordered a bottle of wine. Lukas turned his head a bit at that, but he could see where they were coming from. We hadn't even got to Base Camp yet, and already the weather was flexing its muscles.

• • •

The first thing you notice in Lhasa is the cameras that are noticing you. They're everywhere. There's isn't a street corner without a stack of them, swivelling around. The cameras are large and obvious, a show of strength in themselves. On all the major thoroughfares and large squares, there are floodlights. And on the roads, traffic is repeatedly funnelled through these goalpost-like structures, forcing traffic to slow down so the facial recognition software can pick everyone up.

And as well as the latest, hi-tech surveillance, there's plenty of the good old-fashioned variety as well. Out on the streets of the capital, you didn't need to be James Bond to know that you were being followed. Everywhere we went was extremely regulated. The China Tibet Mountaineering Association, the Chinese Mountaineering Association in reality, decide which hotels you're going to stay in. There's a Shangri-La in the city, but we were put up in a far more rudimentary place. And rather than use the high-speed trains they'd built, they put us in a load of taxis. At the hotel, you have to hand your passport in when you arrive and get it back when you leave. In the cars, you're not allowed to talk to the drivers, can't ask them to stop and pull over anywhere.

Potala Palace, Tibet.

Lhasa is one of the highest cities in the world. At 3,656 metres high (Kathmandu is 1,400 metres by comparison), spending a night there was useful in terms of acclimatisation – though I suspect the mountaineering association put it on the schedule more to ensure we spent some money in the city. And away from the cameras and surveillance, Lhasa remains an incredibly beautiful and fascinating place to explore. The old town, rather than being torn down and modernised, was in fact protected and preserved. I loved seeing the Tibetans in their colourful jewellery, their turquoise and orange corals. I watched people in traditional dress on horseback making the pilgrimage to Potala Palace. The palace itself is a dzong fortress and was the winter palace of the Dalai Lamas for over 300 years. It is both stunning and dramatic to look at; its red and white brick defences cut steeply into the side of Marpo Ri, or Red Hill.

Like the old town, the palace has been left untouched. We went to visit it, and in different circumstances, it might have been an amazing experience. As it was, with all the surveillance, you could never quite shake the tenseness of the atmosphere. In the evening we went out for a dinner. The Chinese businessman who was going with us as far as Base Camp was friends with the former governor of Tibet. That governor had a track record – he had quite an authoritarian reputation, but we did the polite thing, and accepted the invitation. The meal was quite relaxed – a mixture of hot dishes which were shared around, washed down with Chinese Tsingtao beers. But like in the city itself, that sense of menace was never far away.

The businessman in me could see what the Chinese were doing in Tibet. They operate in terms of thirty-year plans, and here the aim seemed to be to develop this as a tourist destination, first for the rest of China and then perhaps more internationally. If you look at the infrastructure they're building into the place, it's the only thing that makes sense. Tibet's population is tiny – about 4 million people – and they are restricted in their movements. Yet they've built and are building these extraordinary high-speed rail links and six-lane highways: that makes sense somewhere like Guangzhou, on the coast between Shanghai and Xiangzhou, but here the population size doesn't justify it. It can only be there to bring people in.

The travel infrastructure isn't the only investment in the area. The old town in Lhasa may have been protected, but there are skyscrapers and shopping malls not far away. There is 4G phone connection throughout, while each village we passed through had running water, electricity and solar panels. Even the smallest villages were kitted out with schools, hospitals and guest houses. All this construction would be expensive anywhere, but doing it at such high elevations increases the cost still further. The contrast with Nepal is stark – fly back from Tibet and you'll notice the poverty, the dirt roads and the uncollected refuse, the buildings without windows, the carboard ramshackle towns.

We didn't get to travel by high-speed train. Instead we were transported in a convoy of small vans, crammed in and sitting on top of our gear. It took ten hours to get to the first town we were staying in overnight, and then another full day's travelling from there. The hotel we stayed in was poor quality as well: seriously, if you put a dog in there, it would rather sleep outside! But because it had all been organised by the Chinese authorities, there was no possibility of checking out and finding somewhere else to stay.

If the transport and accommodation weren't anything to write home about, the views from the van windows were spectacular. These pristine highways, amazing feats of engineering, took us high up through the mountains. We twisted up and down on S-curves. Such was the steepness of the drop on each side that it wasn't a place to lose control of the car. Everything here seemed of greater magnitude than real life: the mountains were taller, the rivers were wider. And the trees: the Chinese were in the process of planting huge forests of trees, billions of them. It was difficult to get your head around just how many they'd planted.

That was part of a concerted plan. The trees help to stop erosion. They have an effect on the climate too: the trees create more moisture, which in turn makes agriculture easier. Beyond that, in the areas where the trees were still to be planted, the landscape was more barren. To begin with, when we left Lhasa, the fields were full of tractors. It was as though every farm worker had one. When I asked about that I was told that it was a Chinese strategy, to make the local economy look good. But the further away we got from Lhasa, the more the tractors disappeared. Agriculture began to look more like it had done for centuries here, with traditional horse-drawn ploughs.

But for all the beauty and contrast of the scenery, one sight in particular stood out on the Lhasa pass. Of all the mountains scraping the skyline, one towered higher than any of the others, its sharp summit shining clear and pyramid-like. Even from this distance, there was no doubting the power and might of the peak I'd come to climb. For months it had been looming large in my mind: now it was doing so across the landscape as well.

Everest.

First look at Everest.

A THROWAWAY PROBLEM

On 22 March 1987, a barge called the *Mobro 4000* slipped out of Islip, New York, heading for Morehead City, North Carolina. Escorted by a small tugboat called the *Break of Dawn*, the barge was large, flat and piled high with rubbish: 3,168 tonnes of the stuff, to be precise. With landfills filling up in New York, the plan was to bury it in North Carolina, and make money by capturing the methane generated by the rubbish. But things didn't quite go according to plan. Instead, what followed was what US TV anchor Dan Rather described as 'the most watched load of garbage in the memory of man'. Other journalists nicknamed the ship the 'Gar-barge'.

When the *Mobro 4000* arrived at Morehead City, the story was picked up by a local TV station, which raised concerns about the trash on the barge. An environmental officer was despatched, who found medical waste on board, leading to concerns the cargo might be hazardous and toxic. A court order was scrambled together, and the barge was told to leave, garbage and all.

Now being tracked by the US media, the *Mobro 4000* continued its journey, looking for somewhere, anywhere, to dock and a landfill to bury its rubbish. By now, everyone knew the barge was coming: the ship's owners were rung by Alabama authorities to be told they had no space for the rubbish, with dock workers instructed to keep an eye out for the barge. Louisiana officials hand-delivered a letter to the ship's owners, telling them they weren't welcome. Port authorities sent out a boat to guard the barge and tugboat, due to fears the tugboat might cut and run, leaving its unwanted cargo behind.

By now the *Mobro 4000* was infested with flies, its decomposing bales of rubbish dripping with black goo. The ship sailed on. Over the course of five months, it travelled on to Louisiana, Alabama, Mississippi, Florida, New Jersey, the Bahamas, Mexico and Belize. All the while, the rumours of what was on the boat grew ever more outlandish. One story suggested nuclear waste might be involved, another the body of missing gangster Jimmy Hoffa.

The latter wasn't true, but developed out of a kernel of truth. Since the mid-twentieth century, many US cities had stopped collecting commercial waste and left companies to find private haulers to deal with their waste. The job that no one wanted to do often ended up in the hands of organised crime – the so-called 'garbage mobsters' took over waste disposal, using it as a way of laundering dirty money. The fact that in the TV series *The Sopranos* the protagonist Tony Soprano runs a waste management firm is very much grounded in reality.

Perhaps unsurprisingly, the garbage mobsters were less than interested in the environmental concerns of the disposing of their waste. In the case of the *Mobro 4000* waste, this was owned by Salvatore Avellino, capo of the Lucchese crime family, who had run Long Island's waste hauling industry for over a decade. Avellino, who later served time for both conspiracy to murder and racketeering, thought he had a good deal for the barge's rubbish: having been paid $86 a tonne to take the trash, he had an agreement to bury it in landfill for $5 a tonne. As the *Mobro 4000* continued its journey and the bad publicity continued, Avellino's company declared bankruptcy, leaving the waste and the ship in limbo. Eventually, five months after setting out, the barge returned to New York. The rubbish was incinerated in Brooklyn and the ash buried back at the Islip landfill from where it had first set out.

The story of the *Mobro 4000* gripped America in that spring of 1987. For a few short months, the issue of how the nation dealt with its waste was high up the news agenda. The fact that so much of the waste system had ended up in the hands of mobsters said everything about how much of a priority was given to the disposal of rubbish. The fact that so much of it was ending up in landfill, and that those landfill sites were filling up fast, also piqued the national conscience. And though that public interest dwindled as the *Mobro 4000* story came to an end, the issues behind it have continued to grow and grow.

· · ·

In October 2019, I visited the Dharavi slum of Mumbai, made famous as the home of Jamal Malik in the film *Slumdog Millionaire*.

I was there to visit the Sharanam Centre for Girls, an orphanage overseen by a good friend of mine, Karen Doff.

Founded in 1882, Dharavi is home to nearly 1 million people. The population density is more than ten times the rest of Mumbai, but it is probably the most efficient square metre usage in the world in terms of productivity. For many years, Dharavi was subject to multi-billion-dollar redevelopment and gentrification plans aimed at transforming the area into a residential and business hub at the heart of the Indian financial capital.

The slum is a hub for small-scale industries, with approximately 5,000 small businesses specialising in things like pottery, leather working, embroidery and waste recycling, some of them more than a century old. Today, Dharavi is a billion-dollar economy with over 15,000 in-house single-room factories for production.

One of the things that struck me on visiting was the strong sense of community that was evident: this is something that we have lost in our large cities, where it is easy not to even know the names of your neighbours. But as a businessman, I was particularly struck by the hive of economic activity from all the small businesses. And what interested me especially was how many of those were involved in the restoration of goods.

I visited shops which fixed old motors, and others which sold second-hand air conditioners. Nothing really died here; the life cycle of products was much longer than in the Western world. Take a 100-year-old Singer sewing machine, for example. Craftsmen here could easily repair the broken parts to make it work. These skill sets do not exist anywhere else in the world

Sharanam Centre for Girls, Dharavi.

and could transform materials that others would disregard into valuable assets.

My Dharavi trip took me back to my own childhood, and to a time when we were more conscious about waste, and about mending and reusing items where possible. When we went shopping, for example, there wouldn't be a plastic bag in sight: you'd take a roller with a bag and carry your fruit and vegetables back home. There were net bags made of cotton, with items wrapped in newspaper if need be.

Everything was reused: I remember my grandmother telling me how she even used to cut up newspapers to use as toilet paper, telling me that you couldn't let anything go to waste. That was partly a consequence of living through the Second World War, when a 'make do and mend' culture was prevalent as part of the war effort. It was an experience that taught the generations who lived through it the importance of thrift and frugality – a way of living that lingered on well into the next few decades.

But at the same time, these beliefs were gradually eroded away by the rise of a more disposable way of living. In 1955, *Life* magazine ran a famous story headlined 'Throwaway Living'. Accompanied by a photo of a family throwing a mixture of plates, cups, cutlery and dishes in the air, the article declared: 'The objects flying through the air in this picture would take forty hours to clean – except that no housewife need bother. They are all meant to be thrown away after use.' Among the items the article cited that could be thrown away after use were a single-use dog bowl, a 'disposa pan' that 'eliminates scouring of pots after cooking' and a disposable diaper, the existence of which the article suggested was 'one reason for the rise in the US birth rate'.

Since then, the rise of disposable culture has continued to grow. It is only very recently that items such as single-use straws, cutlery and coffee cups have started to decline in popularity, or in more developed countries have been banned. Beyond single-use items, many goods we buy are built with what is known as 'planned obsolescence': they are designed to have a set shelf life in order to encourage the consumer to go and buy a replacement. One of the earliest examples of this was the so-called Phoebus cartel of the 1920s, in which the world's leading electric bulb manufacturers agreed to artificially reduce the lifetime of a bulb to 1,000 hours (to give you an indication of how long a light bulb could last, the 'Centennial Light' at a fire station in Livermore, California has been going strong since 1901). Printer cartridges and mobile phones are modern examples where claims of planned obsolescence are common, with microchips in the former signalling the cartridge is empty before it is, and the batteries in the latter slowing down on older models after a set number of years, to encourage the purchase of a new handset.

Inevitably, as consumption has gone up, so has the amount of waste we produce. In 1994, the Jiangcungou landfill in Xi'an City, China, was opened. It is a huge mega-dump: 700,000 square metres in size, 150 metres deep and with the capacity to store more than 34 million cubic metres of rubbish. The landfill was designed to last until 2044. By the end of 2019, it was already full, twenty-five years ahead of schedule.

A serious rethink about the waste that we produce is clearly needed. That's true both in terms of the amount of waste we produce, and also what we do with it. The more we can get back to the ways of those small businesses in Dharavi – recycling, restoring, renewing – the better it will be for our planet. The good news is that there are already a number of brilliant entrepreneurs out there who have already started this process of rethinking the whole concept of garbage. I'd like to introduce you to some of them now.

● ● ●

One of the largest growth areas in waste in recent years is that of e-waste, the name given to discarded electrical or electronic devices.

In 2019, the world generated 53.6 million metric tonnes (Mt) of e-waste, a growth of 21 per cent over five years. Estimates suggest that this will rise to 74.7 Mt by 2030 – a doubling in a decade and a half. As with some of the other figures we've looked at, these numbers aren't evenly spread across the world: Asia accounts for almost half the quantity of e-waste, though in terms of the amount of e-waste generation per capita, Europe has the highest level.

The problem with e-waste is how little of it is recycled, and how very little is recycled properly. Out of that 2019 figure of 53.6 Mt of e-waste, only 9.3 Mt (17.4 per cent) was recycled. Unlike something like, say, a glass bottle, which is made up of one material, a mobile phone or a laptop is made up of numerous components: metals, plastics, lithium batteries and so forth, which traditional recycling units are not equipped to deal with (a typical device can be made up of fifty or sixty separate elements). The result is that many of these items simply end up in landfill. And the problem with that is that because of some of the components they contain, these can go on to leach hazardous material into the land: it is estimated that 2 per cent of the waste in US landfill is e-waste, but that this accounts for 70 per cent of the toxic waste.

Potentially, however, there is a lot of value in this material. For every million mobile phones recycled, it is possible to recover 35,274 lbs of copper, 772 lbs of silver, 74 lbs of gold and 33 lbs of palladium. According to the US Environmental Protection Agency, about 152 million mobile phones are thrown away in the US each year. In terms of the gold that could be extracted from these phones, that totals $200 million.

Mark Evans is the CEO and one of the founders of Camston Wrather, an award-winning US company who are developing a mixture of what they describe as advanced materials engineering and green chemistry to revolutionise the way we deal with e-waste. Mark grew up in a small mining town called Eagle Mountain. The pit was an open-pit iron ore mine, and Mark worked there when he was young, before becoming an electrician, an electrical engineer, and going on to study at Berkeley, Harvard Law School and MIT. He described how the recycling of electronic waste is an 'our generation event'. Growing up, home computers were 'something you watched on *Twilight Zone* or a movie. This was 1977, 1978. And so what's happened is that it's our generation that has created this waste of electronics with the rise of technology, especially home computing.'

One of the challenges was that traditional recycling wasn't equipped to deal with these new items. 'What happened was that it went to the people who collected junk from the 1940s and 1950s; the scrap metal people in your community all of a sudden became qualified electronics recyclers. People said, we let them take refrigerators and other scrap metal, we might as well let them take our electronics. And therein started a number of problems: a lot of people, the bad apples, did it the wrong way. And that's why only twenty-two out of fifty states have regulations on the books about managing electronic waste.'

The result has been that recycling in the US (and across the world) is often quite a diffuse and fragmented process, rather than large companies being able to deal with the recycling on a scale to make the business profitable. Mark described how a lot of the e-waste is now part of a global quasi-economy, where it is shipped and sold abroad. To get around this, Mark explained how a recycling company needed to be 'first in line' to get the products before they are sold on. That involves the growth of brand recognition and a system of collection points, where consumers can recycle defunct items. 'Our biggest competitor is the dump,' Mark explains.

Once Camston gets hold of the e-waste, their process allows them to separate the materials out into what Mark describes as different streams. 'In our process, we make three pure streams that are ready for the refiner. The first is polymers and plastics. We have a magnetic fraction which contains iron, cobalt, lead and rare earths, ready to go. After that, we have a copper and gold product that we make. We sell that to the smelters. For us, it is truly recycling in the best sense.' The potential, if this can be done to scale, is huge. According to Mark, there is 3,000 times more gold in an average tonne of e-waste compared to the equivalent amount of gold ore.

* * *

'I had the idea of the Plastic Bank when I attended an event in Silicon Valley called Singularity University.' David Katz, the founder of Plastic Bank, explained to me the origin of his amazing idea from his home in Vancouver. 'I was there looking for a solution for marine debris. And this idea came out of manufacturing and 3-D printing. I saw a single strand of plastic being turned into a belt. I asked, "What was the price of the belt?" They sold it for $80. What was the material cost? $10. It was only the shape of the plastic that determined the value. I thought, "How might we change not the external shape, but the internal value?" And from there, I thought, "There is a nucleus to an idea, where we can have plastic become money in the world.'"

Like Mark Evans, David dug down into the idea that rather than seeing waste as worthless, we should instead give it a value. The result was the founding of his Plastic Bank project. Currently running in a number of countries around the world, including Haiti, the Philippines and Indonesia, the scheme involves the setting up of a number of Plastic Bank branches where people can bring plastic they have found to be recycled. The plastic is exchanged for money, or alternatively fresh food and clean water, or school tuition for their children. The Plastic Bank then takes the plastic collected, recycles and processes this into a new raw material called Social Plastic, which is then sold on to manufacturers to produce environmentally and socially ethical products. It's a way of monetising the waste plastic in a way that benefits the local community and also reduces the making of new plastic.

David told me about one of the latest schemes which has just been set up in Egypt. 'Egypt is one of the greatest contributors to marine debris in the Mediterranean. The Nile is one of the top contributing rivers to marine debris in the world. The entire country is in desperate poverty. There's probably 40 million now that live under the international poverty line after COVID. But now, for every mother in every household, every piece of material that comes into her household she doesn't have to put out in the garbage. She can think, "Oh, I can bring that to the plastic bank. And I earn an opportunity, I earn money." What was once garbage in the house is now diamonds.'

As with Mark Evans and the potential value to be accrued from e-waste, so David talks about the potential to be earned from recycling plastic. 'Nearly all the plastic we've ever produced is still here,' he reminded me. 'And we keep making more. Nine billion kilos, or thereabouts. Now, even at a very low value of, say, $0.50 per kilo, that's a $4.5 trillion market opportunity. And the beauty of that material is that it could almost be infinitely recycled.'

David is very modest about his achievements. 'It's so self-evident,' he says. 'It's so simple, it's not innovative at all. Nothing that we need to do in the world is against the laws of physics. It's all freaking available. It's just a change of thinking,' he explains, though adding, 'which for some, seems to be more insurmountable than changing physics.'

• • •

This change of thinking is something that Mark Evans mentioned to me as well. 'I have been to so many conferences where we've run into two different types of people at the larger OEMs [original equipment manufacturers]. You have the folks who are the directors of global sustainability. They have all this wonderful data and these facts. But then you have the operations guy. As soon as they see the sustainability guy walk through their door into their office, they're thinking, "This guy just wants to cost me money and time."'

I know from my own experience what these discussions can be like. For many years, there has been a misconception that consumers don't want goods made out of recycled materials – that somehow they'll see them as inferior. But now, the evidence is there that consumers are voting with their money for such products. Not only are they willing to pay for goods that have recycled materials, they're actually willing to pay more. In an industry like mine, that's been something of a paradigm shift in recent years.

In the manufacture of our washing machines, for example, we've started using PET bottles to make the machine tubs. The original expectation was that this would increase costs: in fact it has reduced them. So far we've used something like 58 million bottles in those tubs. In our ovens, we've started using ghost nets – the name given to nets discarded or cut off by fishing boats and left behind in the ocean. Eight tonnes of fishing nets and ninety-nine tonnes of textile waste were recycled into materials to be used in products. Again, the process makes money, because it reduces the amount of virgin materials needed to make the product.

One of the problems with waste is that for too long it has been left in the hands of the consumer to deal with. But it is time for all manufacturers to step up and take more responsibility. Two ways that I've personally tried to tackle this are by taking on the return of our products and changing the packaging of our goods.

In Turkey, we've built two giant recycling facilities ourselves and have started recycling all the appliances that our customers no longer want: fridges, washing machines, dishwashers, ovens, basically every large appliance we make. To date we've recycled more than 1.3 million appliances, which in our industry is a huge number. In the beginning, we were giving those component parts to people who sell recycled materials, and they were selling it to industries downstream. But now we've started using it ourselves, and we use a big chunk of it back in our own products.

We're also looking at ways that we make all our packaging material recyclable. We're working to remove all Styrofoam, which I think is one of the biggest offenders in terms of plastic pollution: it lasts forever and breaks down into lots of small parts. We've replaced that with a sort of recycled paper in all our small domestic appliances already. And we've also removed the colourful prints on the boxes, because whenever boxes have that glossy outer coating, the toxic materials involved in the printing make them non-recyclable.

These are just two examples of ways that I've managed to change things in the company that I run. But whatever sort of business you might be involved in, there are always ways that you can make a difference. As David Katz says, the key is to think differently. And when you do, you'll be surprised about the economic as well as the environmental benefits of your actions.

BASE CAMP AND ACCLIMATISATION

Base Camp is a bit like landing on the moon. Or maybe, given the colour of the rock as the sun comes up, Mars. It's a flat, stony plain, 5,000 metres high, with nothing there. There are no animals, no birds, no insects, no bugs. You drop a bit of food and it's there the next day, exactly as you left it.

Our camp was made up of what at first looked like rows of orange sheds. These were actually tents, but bigger and more solid than what we'd be sleeping in further up the mountain. There was a bed to lie on, by which I mean a frame and mattress to put your sleeping bag. That might not sound like much, but in mountaineering terms, it was luxurious. There was electricity, so you could charge your phone, and a light bulb. Again, that might not sound a lot, but it made a big difference. There was even a desk and a chair. I really appreciated the latter, as it made such a difference taking your boots on and off.

Everest Base Camp, Tibet.

As well as the rows of orange sleeping tents, there was a smattering of larger, white domed tents. These were communal tents. One was an entertainment tent, where you could watch TV or listen to music. There was a table and couches too, and various books that people had left behind that you could read. There was also a dining tent, with a large table for everyone to sit around. Even better than the food, this tent, unlike the ones we were sleeping in, had heaters.

They weren't the best heaters, just gas heaters that went on and off, but compared to the cold outside, they seemed amazing. The rise and fall of the temperature at camp was remarkable. During the day, if the sun was out, it could get really hot; there's nothing alive around you, and the heat just bounces off the rocks. But at night, the temperature would plummet – it would go down to minus thirty.

We all had pee bottles to urinate in. You'd hang these outside and almost immediately, I noticed that everyone else's had a larger mouth than mine. I was a bit embarrassed about that – was there something going on that I didn't know about? Would people be making comparisons? But then I discovered that with the low temperatures, when you urinated at night, it froze almost straight away. If you had a large bottle with a big mouth, then the ice block would fall out. If you had a bottle with a small mouth, like mine, then the ice would stay there. It was one of those mistakes, I learned, that a lot of rookie mountain climbers make.

For all of the ways of relaxing offered at Base Camp, it was the view of Everest that I found myself watching. We were properly close now, unlike when I'd seen it on the Lhasa pass. Here, the mountain was sharper, more in focus, more dominating. From Base Camp, it seemed crazy that we were going to attempt to climb it. It looked too high, too steep, too impossible.

I watched as the weather weaved its way against it, the shadows shifting as the sun moved across the sky, the small explosions of snow as a gust of wind caught the side, the plumes of cloud catching across the summit, then pulling free to reveal the peak, hanging off the side as though the top was smoking. From barren rock where we were, the tints of yellowish stone gave way to the greys of the higher slopes, scored with snow to give a marble effect. Behind, the sky was a clear, cobalt blue, bringing the mountain into relief still further.

At one end of the Base Camp, with Everest looming in the background, there sits a small collection of memorials, homage to some of those brave souls who had lost their lives on the mountain slopes. They're simple affairs, in keeping with the surroundings: cairns of stones piled up, interspersed with a smattering of more formal engravings. One granite-grey slab of marble in particular caught my eye. It was nestled into a mound of rocks – stones left by climbers in memory of one of the mountain's most famous sons.

In memory of George Leigh-Mallory and Andrew Irvine
Last Seen 8th June 1924
And all those who died during the pioneer
British Mt Everest Expedition

When my imagination was first fired by Everest, back when I was living in Hong Kong, I'd read round about some of the most famous expeditions. At the time, Jon Krakauer's *Into Thin Air* had just been published. This was the account of his ascent of Everest in May 1996, when eight climbers died after a blizzard hit the mountain on the descent. Beck Weathers, who I'd invited to speak in Istanbul, was one of the other survivors of this particularly deadly episode (the events were later turned into the 2015 film *Everest*, where Josh Brolin played Weathers and Michael Kelly the role of Krakauer). Coupled with four other deaths on the mountain that season, it left 1996 as the deadliest in Everest's history (this tragic total has since been overtaken by the 2014 and 2015 seasons, when avalanches resulted in sixteen and twenty-two fatalities respectively).

I was one of many people who became captivated by Krakauer's story. You'd have thought that story of so many people dying on an expedition might put me off climbing Everest for life, but instead it stirred me to investigate the mountain's history still further. I had a New Zealand girlfriend at the time, and partly through her, I became interested in Sir Edmund Hillary. It was he, together, with Sherpa Tenzing Norgay, who became the first people to scale the summit on 29 May 1953. But while Hillary and Tenzing succeeded where others had failed, it was the pioneering work of previous expeditions on which their success was built. One of those mountaineers whose journey to the summit remains a mystery was George Leigh Mallory, who lost his life in 1924 during his attempt to conquer the mountain.

Mallory was someone I both aspired to be (a brilliant moun-taineer and dashingly handsome) and could relate to (he was married with children). Following an early career as a teacher (the poet Robert Graves was one of his students) and fighting with distinction during the First World War (including at the Battle of the Somme), he then turned his attention to mountaineering.

When asked by the *New York Times* why he wanted to climb Everest, Mallory famously replied simply, 'Because it's there.' In 1921, he was part of the British reconnaissance expedition to explore if a route up the mountain was possible. It was sixty-five years since a previous mission, the Great Trigonometrical Survey of India, had measured what they concluded was the tallest moun-tain on earth. That survey had begun much earlier, back in 1808, tasked with mapping the Indian subcontinent. By the late 1840s, the mission had reached the foothills of the Himalayas. At the time, Nepal was closed to outsiders, and so measurements were taken from approximately 100 miles away.

It was down to the brilliance of Radhanath Sikdar, a Bengali mathematician, that the height of what the survey called Himalayan Peak XV was calculated. 'Sir, I have discovered the highest mountain in the world,' Sikdar reportedly announced to the surveyor general. Up to that point, Kangchenjunga had been assumed to be the world's highest peak. Everest was named after Sir George Everest, leader of the survey, with both the Tibetan name (Chomolungma) and a Nepalese one (Sagarmatha) being unknown to the survey. Chomolungma, depending on which account you read, means either 'Mother Goddess of the World' or

'The Mountain So High No Bird Can Fly Over It.' George Everest's name, by the way, is actually pronounced Eev-erest, though that also got lost somewhere along the way.

Given the comparatively rudimentary tools at his disposal, plus the restrictions of distance, it seems remarkable that Sikdar's calculations were within ten metres of the mountain's actual height. Sikdar's estimation was that Everest was 8,839 metres; it wasn't until a century later that an Indian survey recalculated it at 8,848 metres. For many years, the Chinese offered a different measurement, suggesting that the mountain was 8,844 metres tall. In late 2020, after a discussion over whether or not to include the snow cap on top, Nepal and China agreed a revised height of 8,848.86 metres.

Although Everest had been measured, it was to be many more years before an expedition was launched to climb it. As with Nepal, Tibet had also refused entry to foreigners until the early 1900s. That changed with the 1904 Treaty of Lhasa, but it was not until after the First World War that an ascent of the mountain was properly considered.

The 1921 expedition team that Mallory was part of had a somewhat different experience of getting to the mountain than I had. Rather than the luxury of being driven there on newly built six-lane highways, they had to walk 500 kilometres across from Darjeeling, often on land that hadn't even been mapped, just to get to the foot of the mountain. From there they began to investigate possible routes up the mountain via the Rongbuk Glacier, with their survey still shaping the routes used today. The highest point reached on the first expedition was North Col. Bad weather precluded them from climbing further, but for the first time, a possible route to the summit could be seen.

The journey to North Col had taken four months. Mallory wrote to his wife, saying, 'I wouldn't go again next year … for all the gold in Arabia.' But in 1922, he was back. This time he got further, in a three-man team alongside Teddy Norton and Howard Somervell, making it to 8,225 metres without the use of supplemental oxygen. Two further members of the expedition team, Geoffrey Bruce and George Finch, did use oxygen ('English Air', as the British teams called it) and got closer still, reaching 8,300 metres, just 500 metres short of the summit. But for all the success of the climb, it was overshadowed by tragedy, with seven Sherpas being killed in an avalanche near North Col.

Mallory returned for his third and final attempt in 1924. This time, his climbing partner was Andrew 'Sandy' Irvine, and this time he was planning on using oxygen. 'I can't see myself coming down defeated,' he wrote to his wife. 'I don't expect to come back,' he wrote to a friend. The expedition leader, E. F. Norton, first made it to 8,570 metres, without the use of oxygen. Then it was Mallory and Irvine's turn to work their way up from Advanced Base Camp, up North Col and on through what are now the higher camps. On 8 June, expedition member Noel Odell spotted the pair from his position lower down the slopes. 'There was a sudden clearing of the atmosphere,' he later recalled, 'and the entire summit ridge and final peak of Everest were unveiled. My eyes became fixed on one tiny black spot silhouetted on a snow crest beneath a rock in the ridge; the black spot moved. Another black spot became apparent and moved up the snow to join the other … Then the whole fascinating vision vanished, enveloped in cloud once more.'

Odell's sighting was the last time that Mallory and Irvine were seen alive. Quite what happened to the pair remains a mystery lost in the mists of the mountain. Odell believed he saw the pair on the Second Step, though no definitive proof was ever found that they climbed this far. A used oxygen cylinder was later found lower down, near the First Step; in 1933 Irvine's ice axe was found nearby. In 1975, a Chinese climber, Wang Hungbao, saw what he described as 'an old English dead' in a gully near the ice axe. He died in an avalanche the following day, before his sighting could be pinpointed. It wasn't until 1999 that another expedition finally found Mallory's body, frozen and preserved (Irvine's body has never been found). There were rope injuries around his chest, suggesting he had slipped, and a wound on his forehead, possibly from hitting his head on a rock or his own ice axe, which was the probable cause of death.

The question of whether the pair made it to the top is unresolved. One possible clue is that Mallory was known to have a photograph of his wife, Ruth, with him, which he intended to put on the peak when he summitted. This was not found among his possessions. The only way left to properly resolve the mystery is if Mallory's camera could somehow be found: that also wasn't among Mallory's possessions. If the camera was discovered, there is a good chance that Everest's frozen, dry conditions would have kept the film preserved, and the truth about what happened almost 100 years ago would finally be revealed. (In 2019, photographer and climber Renan Öztürk set out to try and find Mallory's camera. His efforts were charted in the 2020 documentary film, *Lost on Everest*. It was fascinating for me to watch, especially with the struggles that experienced climbers' faced on the mountain. That helped put my own challenges into perspective.)

• • •

As soon as we arrived, Lukas started putting us through our paces. For all the sleeping in my hypoxic tent, there was still plenty of acclimatisation to do. That meant a run of 6,000-metre climbs. Aconcagua as a climb had been just under 7,000 metres tall: now we were doing 6,000-metre climbs as training, one trek after another. Lukas wanted us to be ready to go as quickly as possible. With the weather window continually shifting, flexibility was everything. It was strange to go from all the hanging around of the previous week to suddenly be on a moment's notice.

The first acclimatisation climb we attempted was a disaster. When I did Aconcagua, the acclimatisation had been a problem, but the climbing itself was OK. I knew that I was physically fit and could take what the mountain threw at me. This time round, I was exhausted. The first stretch was fourteen miles from Base Camp to the interim camp, and I wasn't sure if I was going to make it. The route takes you through a deep valley, where you have to climb down through loose rock, cross a stream and then go back up the other side. I remember looking down at the path and thinking, there's no way I can get back up on the other side to get to camp. I was by myself at this point, having been left behind by the others. I was reduced to taking it one step at a time, talking myself through it.

Each step, though, was a sear of pain. I'd done everything I was told to prevent blisters – put the plasters on early, wear a second pair of thin socks – but here I was on day one, and my feet were a mess. I could feel the blisters every time I put a foot down. I'm normally an optimistic person, but that really got me down. This is the first day, I thought. If I'm tired now, how am I going to make it to the top? And if I'm getting blisters already, how are my feet ever going to heal?

When I finally made it to camp, I went straight to my tent and collapsed. I really needed to go to the loo but didn't have the energy to move. My feet were throbbing. My mouth was completely dry and I had telltale signs of problems with breathing. And then, exactly as at Aconcagua, I couldn't sleep. At least on Aconcagua, because I was fit, I wasn't physically tired. Now, though, I was properly exhausted, and needed that recovery time. My only consolation was that it wasn't just me who was feeling like this. Fred and Marco were completely destroyed as well. Jann Rageth, the Swiss businessman in the group, had only gotten to sleep by taking sleeping pills.

There were also comforting words from another member of the team, who'd just come back down from higher up the mountain.

'Don't worry,' he told me. 'The second time around it gets much easier.'

'I don't think there's going to be a second time round,' I said. 'I'm heading back down this mountain and not coming back again.' The team member gave me a nod, like he'd heard it all before.

The second day, however, was no better. We climbed another eleven miles and I was in pieces. Lukas took one look at me and told me to take a sleeping pill. This time, I managed to sleep for a couple of hours, which made all the difference. The team member who'd come back down the mountain was right: it's amazing how quickly your body adapts. I was still tired and still challenged, but I wasn't the wreck I was after the first day.

When we made it back down to Base Camp, Fred and Marco took off immediately. Rather than waiting around at Base Camp, they headed for the nearest town to recover. I decided to stay at camp. I'd come all this way and felt I should be there. I spent a couple of days getting to know Base Camp a bit more. Everyone was in their separate sections, with their own different way of doing things. There was a Russian camp which had a more relaxed attitude to preparations. They had a massage tent, and another with a bar containing hundreds of bottles of vodka. The Russians held camp parties and invited everyone along. It was surreal. Their success rate was much lower than Lukas's, but everyone had a great time.

There were German members of our expedition, too, who had been at Base Camp much longer, rather than being on our rapid acclimatisation programme. I remember a female member of the group who had this hacking cough. When I stepped away from her at meal time, she told me that she had the 'Khumbu cough'.

'Don't worry,' she told me. 'I'm a doctor. This isn't contagious.'

But I kept my distance all the same.

After a couple of days, I decided to join Fred and Marco. It wasn't cheap, booking a lift back down, but it was worth it. The town had a Burger King, and I'm afraid to say I tucked into a Chicken Royale. It was disgusting and deeply satisfying at the same time. I finally had a reasonable cup of coffee, a night in a half-decent bed, and felt much better.

All the time we were down there, we were in communication with Lukas's team. They were constantly monitoring the weather to figure out when the best window to attempt the climb would be. When word came through of a possible early weather window we scrambled back up to camp. This time, although the forecast was OK, we were hampered by Chinese bureaucracy.

There had been reports in the national media about the condition of Base Camp as a result of global warming, with rubbish and sewage and dead bodies turning up because of the melting of the glacier. As a result, the Chinese had a sent an official from Beijing to investigate. The problem with that was that further up the mountain, Chinese officials had been fixing that season's ropes. These were the ropes that everyone used to climb up, and they were carefully checked and refreshed where necessary. But with the Beijing official on his way to Base Camp, the rope layers were called down from the mountain to meet him. So while the weather was OK, the ropes weren't ready.

As the days ticked on, missing the window became increasingly frustrating. Keeping track of the weather was a crucial part of the expedition planning. So while myself, Fred and Marco dropped back down to the nearest town to wait it out, Lukas and his team were continuing to pore over the data that the weather forecasters were offering. The shift in quality of those forecasts over the last few decades has been remarkable. This is thanks to a huge increase in processing power: the Met Office's supercomputer now has the capacity to do over 14,000 trillion calculations every second. The computers they were using thirty years ago, by comparison, had the same power as a smartphone.

What this extra processing power means in practice is that the weather can be analysed in increasingly smaller grid squares, making weather reports more accurate. This is particularly crucial in forecasting mountain weather. On Everest, the particular geography and altitude means that there are a lot of variables in play. Wind behaves differently on steeper slopes, and can be far stronger the higher up you go. Then there are the shifts in temperature: at altitudes, these can shift more quickly and more dramatically. And there are cloud movements to consider as well, crucial for visibility and with the potential for snowfall. It's a lot of number crunching to produce the models and forecasts needed to work out the optimum time to summit.

I was climbing Everest to raise awareness about climate change, but climate change itself was having its own effect on Everest, with weather patterns becoming more unpredictable and more extreme. At the start of May, a cyclone had moved in from the Bay of Bengal to hit the Indian eastern coastal state of Odisha. The Indian Meteorological Department had described Cyclone Fani as 'extremely severe' and rated it a Grade 5 storm. It spent ten days gathering strength over the seam before hitting the coast with winds of up to 127 miles per hour.

Over a million people were evacuated, but the storm's reach spread far beyond Odisha: its impact could be felt in dust storms in Rajasthan, heatwaves in Maharashtra, torrential rain in the Chinese borders and heavy snow and high winds in the Himalayas.

On Everest, those already on the mountain making preparations for the coming climbing season had to descend back down the mountain. Tents that had been set up in the higher camps were blown off, twisting and ripping in the wind before being tossed away.

The human cost of Cyclone Fani was less than that of 1999's Cyclone Odisha. This was thanks to those weather supercomputers, and the better forecasts they were able to offer. But that extra processing was sorely needed. Cyclones don't usually happen at this time of year in the region: the majority of cyclones to hit India usually take place in the post-monsoon season between October and December. For a cyclone of this size to hit in May was a telling reminder of just how much weather patterns were shifting. Talking to those on the mountain, it was clear how unusual an event this was, which made my ascent of Everest feel all the timelier.

Back down in Tingri County, I was initially worried that we were losing acclimatisation. But then we met up with two other climbers, Cory Richards and Esteban 'Topo' Mena, two of the most experienced mountaineers in the world, who had both summitted Everest several times. If they were back down here, then it must have been OK!

It was an amazing opportunity to speak to them, and get a sense of what it took to become an elite climber. Richards and Mena hadn't just climbed Everest before – they'd both done so without oxygen. Richards, a *National Geographic* photographer, made his mark in the mountaineering world by making the first winter ascent of Gasherbrum II (another of the world's fourteen 8,000-metre climbs). Mena, who comes from Ecuador, had scaled the south face of Aconcagua (a different league of difficulty to the route we took) when he was just nineteen. They had returned to Everest to try and climb its north-east face, on a new route never attempted before – and the first attempted addition to the mountain's eighteen recognised routes for more than a decade.

For those into climbing, Richards and Mena were a different type of rock star. I felt enthused and reassured by our meeting, and grateful for the advice that they gave us. So when the call came from Lukas for us to return back to Base Camp, I was pumped to get going. The weather forecasts had predicted a two-day window when it was possible to get to the summit, so there was no time for hanging about.

The ascent was on.

FOOD FOR THOUGHT

'There was once a town in the heart of America where all life seemed to live in harmony with its surroundings.' So begins 'A Fable for Tomorrow', the opening chapter of one of the most influential environmental books of all time. Its author, Rachel Carson, describes a fictional American town, once rich with nature, but where 'a strange blight crept over the area and everything began to change … everywhere was a shadow of death.'

Most ominous was the eerie silence that now cloaked the land: 'The birds … where had they gone? Many people spoke of them, puzzled and disturbed. The feeding stations in the backyards were deserted. The few birds seen anywhere were moribund: they trembled violently and could not fly. It was a spring without voices. On the mornings that had once throbbed with the dawn chorus of robins, catbirds, doves, jays, wrens and scores of other bird voices there was now no sound.'

The American town portrayed in the opening of *Silent Spring* might not have existed, but all the components and natural disasters Carson had pulled together had occurred in one place or another. She had been inspired to write the book after a letter from a friend in Massachusetts, describing the mass death of birds following the spraying of DDT. DDT was, at the time, the most powerful pesticide in the world. During the Second World War, it had been developed by the Allies and used to rid South Pacific islands of malaria-carrying insects, for which it was extremely effective: the only problem with it was that rather killing the insects specifically, it also obliterated hundreds of different species at the same time.

Meanwhile, on the other side of the war, the Germans had been developing organophosphate nerve agents, so that they could be used as chemical weapons. While that plan never reached fruition, the technology was appropriated by American companies after the war, for use in agriculture in the making of artificial fertiliser. Fresh in American minds was the experience of the 1930s, when the Dust Bowl led to drought, economic misery and the devastation of agricultural lands. With a new array of pesticides and fertiliser at their disposal, and helped by government subsidies to farmers, agriculture flourished. The American food supply was secure, reliable, and, using new methods, producing food that was cheaper for consumers than ever before.

But this abundance came at a cost, as the letter Rachel Carson received in 1958 told. Carson, a bestselling nature writer, decided to investigate. Having failed to persuade any magazine to commission her to write a piece on the subject, she wrote a book about it instead to raise awareness. When it was published in 1962, it was strongly attacked by the chemical industry. Argo-chemical giant Monsanto published 'The Desolate Year', a parody of Carson's 'A Fable for Tomorrow' chapter, depicting an America without pesticides: 'The bugs were everywhere. Unseen. Unheard. Unbelievably universal … beneath the ground, beneath the waters, on and in limbs and twigs and stalks, under rocks, inside trees and animals and other insects – and, yes, inside man.'

Carson, though, was to be vindicated. The then US president, John F. Kennedy, asked his Science Advisory Committee to investigate the claims in *Silent Spring*. The scientists came down on Carson's side, leading to the banning of DDT and tighter regulations regarding pesticides and fertilisers more generally.

Despite such efforts, we're now living in the world of Carson's tomorrow, and her fable continues to be sadly prescient.

Spring 2020, and the waters off the southern coast of America in the Gulf of Mexico turn a strange shade of green. This is algae, which in the natural cycle of life grows, dies and sinks to the bottom of the sea, where it is decomposed by bacteria. But over the last few decades, the amount of algae growing in these waters has ballooned. And when it sinks to the bottom to decompose, the bacteria consume oxygen to do so. The more algae, the more decomposition the more oxygen used up. The result is a hypoxic or anoxic area (low or oxygen-free). Or, to give the area its colloquial term: the dead zone.

Without oxygen, no fish can survive. The creatures that can swim further and further from shore stay alive; those that can't, perish. Quite how far the surviving fish go you can track by the journeys of local fishermen. Each year, they have to sail further and further out to sea in order to make their catch. Over the last five years, the average size of the dead zone has been 5,407 square miles (in 2017, it stretched out to 8,776 square miles).

Why the surge in growth in algae? It's because it is feeding on nutrients. Where do the nutrients come from? From fertiliser and other chemical discharges from further up the Mississippi. This is deposited in the water from the Midwest, the Corn Belt states that are the home of American agriculture, and then washed down into the ocean. In May 2020 alone, to give an idea of scale, the NOAA (National Oceanic and Atmospheric Administration) estimated that 136,000 metric tonnes of nitrate and 21,400 metric tonnes of phosphorus were discharged into the Gulf Mexico.

'A grim spectre has crept upon us almost unnoticed,' Rachel Carson wrote in 1962, 'and this imagined tragedy may easily become a stark reality we all shall know.'

* * *

The effect that agriculture has on climate change is easily overlooked. Yet research shows that the livestock sector – meat and dairy – accounts for 14.5 per cent of greenhouse gas emissions. This is more than the emissions created by transport on land, sea and air: all cars, trains, ships and aeroplanes combined. It is more emissions into the atmosphere than the US economy. And what's more, estimates of meat and dairy consumption are projected to continue growing over the next forty years: meat by 76 per cent

Greenhouse gas emissions per kilogram of food product

Greenhouse gas emissions are measured in kilograms of carbon dioxide equivalents (kgCO₂eq) per kilogram of food product. This means non-CO₂ greenhouse gases are included and weighted by their relative warming impact.

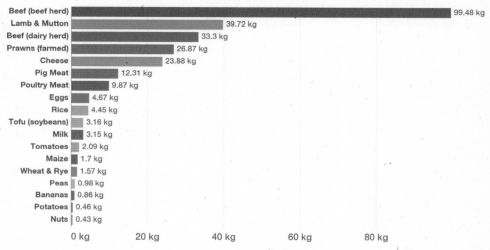

Source: Poore, J., & Nemecek, T. (2018). Reducing food's environmental impacts through producers and consumers.
Note: Data represents the global average greenhouse gas emissions from food products based on a large meta-analysis of food production covering 38,700 commercially viable farms in 119 countries.
OurWorldInData.org/environmental-impacts-of-food • CC BY

and dairy by 65 per cent (in Asia, it is projected that meat demand will rise by 300 per cent).

These emissions are made up of two key factors. Nitrous oxide comes from manure and fertilisers used in the production of feed. Methane comes from what is politely described as 'enteric fermentation', or in language your children would understand, cows burping and farting it out. In 2019, researchers at the University of California estimated that a single cow on average belches out 100 kg of methane. If you want to know how much methane that produces in total, in 2019, the global cattle population was estimated to be just under 1 billion. On current trends these emissions are anticipated to double from 1995 levels by 2055.

In terms of maths, the equations here are fairly straight-forward. As the global population rises, so does the demand for food. Unchecked, the production of meat and dairy will continue to grow, creating additional levels of greenhouse gases in the process. To bring these levels down to a sustainable level, we need to change our diets and modify our consumption of these food sources. That, though, is easier said than done, as anyone who has ever been on a diet can testify. And even if people can be persuaded to change, what should we be eating instead?

· · ·

'If tomorrow everybody just stopped eating meat, it would be a very quick way to reduce a lot of short-lived methane – a potent greenhouse gas – in the atmosphere. Philosophically or theoretically, it's the fastest thing we can do to mitigate climate change.'

Christiana Musk is one of the world's foremost thinkers and campaigners on food and the environment. She is the founder of Flourish*ink, a platform for conversations on the future, as well as *The Future of Meat* podcast. She is also curator of food content for the Near Future Summit and the former Executive Director of Food Choice Taskforce. Christiana also worked alongside Chatham House to develop the award-winning research project, Livestock – Climate Change's Forgotten Sector, which revealed the extent of environmental damage caused by this sector. When I met with her via Zoom during the pandemic, she told me about the project and its findings.

'It was the largest global study on meat consumption and climate change,' she explained. 'We held stakeholders meetings on reducing meat consumption for climate change in China, Brazil, the US and the UK. We did surveys in seventeen countries and looked at consumer willingness to change their diet, at climate awareness around meat consumption and climate change.'

What the report discovered was a large variation from country to country as to the willingness of people to change their diet. 'It is very difficult for people to consider their individual role. I worked with engaging a lot of different environmental organisations around this. There's a lot of people with local expertise on all the spectrums of interventions to try to get people to change their diet. But what it comes down to is that it's probably not going to happen fast enough.'

I asked Christiana why this was the case. 'We've found that just doing it for climate is not a motivator enough. People say that health is a motivator enough, but when we work with the public health folks on this, people find it really hard to change their behaviour even to save their own lives. Even if they have a doctor saying, "You're going to die of diabetes, and a stroke, and hypertension." We have 2 billion who are suffering from lifestyle diseases, and they could prevent their own deaths and suffering, and they just can't stop eating the food. It's easy to write in a book, "Yes, it would be great if people changed their diet." The question is, how do you really motivate people to do that?'

Christiana explained how her focus turned to different forms of persuasion to encourage change. 'What I became more passionate about was looking at ways of shifting the environments in which people are making choices. To give you an example, a lot of people eat food that's curated for them by an institution. They work in a government building and there's a cafeteria; they go to a school and the government has chosen what the school diet is. Procurement is very significant, a big player in deciding what are the choices people have to choose from. If you make commitments to buy things that are produced with environmentally friendly operations, and are low carbon, then everything people have to choose from is climate friendly.

'Secondly, you can play with the portions. There's some great research about how you can play with choice psychology by the way that you set up the cafeteria. Usually cafeterias have your meat first, and so you're like, "Oh, which protein do I want? And then I'll tuck some veggies on the side of it." But if you put all the vegetable and high-fibre things first, people fill their plates with that and don't have that much room for the meat. Then you can put educational materials around to encourage people to say, "Hey, did you know how many calories this has in it?" How much land you're saving, or how many animals. When you combine all those things together to make it really easy for people, a lot of times without people even noticing, they change their choices.'

• • •

I asked Christiana about the role that technology could play in offering solutions. She described the industry as 'a messy sector compared to other sectors that can have technological solutions, because food is an exchange with our bodies. And also so many rural livelihoods still depend on the production of agriculture. It's hard to have a straight up, "We're going to create a new product and it's going to solve this, and it's going to be more efficient."'

One solution touted is the creation of meat substitutes. In recent years, the market has seen the launch of products such as Impossible Foods and Beyond Meat offering plant-based meat substitutes that claim to taste the same as the real thing. But although there is no meat involved, these products can still end up being highly processed. 'I started to question the healthiness of some of the new generation of vegan burgers,' food writer Bee Wilson wrote in the *Guardian* in 2019, 'after I ate a Beyond Burger, as served at the Honest Burger chain. While eating the burger … I was stunned by how close it felt to meat in my mouth, with its rosy pink hue and fragile flesh-like texture. But it felt nothing like meat to my digestive system. Half an hour after lunch I started to have griping stomach pains and a horrible junk-food aftertaste. When I looked up the ingredients, it occurred to me that had they not been marketed as quasi-meat I would never have chosen to lunch on "pea protein isolate, expeller-pressed canola oil, refined coconut oil, water, yeast extract, maltodextrin, natural flavours, gum Arabic …"'

Christiana suggests that there is still a way to go in this field: 'Everything that you see today in the media is very, very early stage. A lot of the companies that I've been working with are going to be much more scalable, much more affordable, less processed and much more acceptable for people. One example is a steak that's mycelium-based: that's the root system of mushrooms, of fungi. The makers are able to brew it in beer tanks. And so they can meet the world's demand for meat pretty easily using a fermentation process. They only use beets as a food colouring, as the source of their red food for the steak, and then flavours, natural flavouring … a lot of people who don't want the highly processed Impossible Burger or Beyond Meat burger will feel comfortable eating mycelium that's been brewed, because it's a type of biotechnology that's ancient and very familiar to us.'

Another possible technological route is the development of genetically modified (GM) crops. These are foods where the DNA has been modified. While GM crops have been developed since the late 1980s and early 1990s, the process has the potential to be revolutionised by the more recent advent of something called CRISPR, which is a technology used to edit genes in living cells. First developed in 2012, CRISPR has since been used in scientific research, but the technology and techniques are expected to be used widely in food production over the coming years.

'I think there's no stopping that we will have genetically modified foods in the future, and lots of them,' Christiana told me. 'I think the technologies are going to be very, very different than today because of CRISPR. We're going to have a lot more innovation very quickly in genetically modified foods. The question is going to be making sure we have proper labelling and communications for people, because of a lot of people would like to avoid those foods.'

The challenge and opportunity with this technology is what the crops are being modified for. A lot of previous development of genetically modified crops has been to improve the yield of 'monoculture' crops (monoculture farming is about growing a single crop), rather than in exploring variety and biodiversity. 'The argument that I make around this is that we have 10,000 years of research and development in crop diversity,' Christiana explained, 'and we have 80,000 species of plants and animals that humans have spent 10,000 years domesticating. And yet, 95 per cent of the foods, the calories that we consume globally, come from ten species of plants and animals. That's an industrial mindset that came out of the productionism in the 1940s and then spread all over the world.

'We don't want to lose all the beautiful genetic variety that we already have. The more that we can be investing in agro-biodiversity at the same time as we're innovating in genetic crops, then we're going to have a lot more food. And I think the future is a lot more colourful, with a lot more variety, a lot more appropriate crops for the region. And when you have more appropriate crops for a shifting climate, and we're able to shift more, then we'll just have a much more beautiful variety of food everywhere in the world.'

Christiana's comments about the possibility for variety echoes a conversation I had with Tom Chi, the man behind the trillion trees initiative that we'll explore in a later chapter. Another project Tom is involved in is Iron Ox, an autonomous robot farm that uses artificial intelligence to help grow plants. 'Think Dutch greenhouses but with only robots working there,' Tom explained. The plants are grown in a hydroponic facility where they are housed in grow modules and monitored and tended to by robots. The set-up is far more efficient than traditional farm activity – up to 95 per cent less water used and thirty times more space-efficient – and more cost-effective too: 'we are already cheaper than growing outdoors, which is the gold standard worldwide for how cheap you can grow,' Tom told me. 'We're on trajectory to end up about three times cheaper than growing outdoors.' The labour costs on the farm are replaced by a robotic energy cost which Tom describes as 'roughly equivalent to the cost of running a lightbulb.'

On top of this, the process requires, 'no pesticides, no herbicides, no fungicides. It turns out that if you use robotic labour, then you can replace chemistry with labour. What I mean by that,' Tom explains, 'is the reason that we spray everything is not because no bug ever bites a thing that's been sprayed. We spray everything to prevent outbreaks. What happens with our plants in the Iron Ox setting is every single plant gets looked at by a robot every single day or every other day. So when you see a single leaf start to get some aphids or you see it getting powdery mildew or yellowing a bit, you can just address it by composting the diseased bit. In the worst case, you compost that whole plant and stop such outbreaks with robotic labour instead of poisonous chemistry.'

Not only does this process reduce the use of chemicals in farming, the ability to focus on each individual plant opens the door for more variety. 'We have different care procedures for our basil,' Tom explains, 'and we can change the intensity of aromatic oils in basil depending on how it's cared for. It can be exactly the same species and variety of basil, but the robot tending it a little bit differently makes it so that the resulting aromatic oils are bolder, more even, upfront, or delayed. Our sense from working with 40-plus crops so far is that pretty much every plant on the planet has a wide range of expression similar to wine grapes. In wine grapes, we have a set-up where individual humans would take care of individual vines. So in a situation where humans expend a lot of labour per plant, we can get a lot of flavour expression. But what is true of these grapes is likely true for most plants. We just have not expressed them as such because labour is expensive. Robotic labour makes this tending for flavour into a modified care algorithm all while significantly reducing the cost of care.'

• • •

It's clear talking to Christiana and Tom that there is the potential for technology and science to be harnessed in ways that can change our diets for the better, and for the benefit of the planet as well. But I also wanted to touch base with someone about the medical implications of reducing the amount of meat in our diets. To do so, I got in touch with Brian Clement, a leading expert on diet and healthcare. Brian is the director of the Hippocrates Health Institute in Florida and author of numerous books on health, spirituality and natural healing, including the bestselling *Living Foods for Optimum Health*. When I first met Stephanie, she put me on a mission to get healthy, which is how I got to know Brian. Brian told me how, historically, we lost touch with the way our ancestors originally used to eat. 'Up until the mid-nineteenth century, nobody was ever told our ancestors were hunters and gatherers. That was never in writing. Anthropology never addressed that. And then a group at Oxford and Cambridge pretty much sat down and thought, "This is the theory we now have.

This is the one we're going to embrace." And that theory hasn't really changed until recently.

'Finally you have some anthropologists who got off their ass and did some work. Somebody started to say, "Let's look at the teeth. Let's look at the indentations on the teeth of Homo sapiens, your ancestors, going back from pre-Homo sapiens to right now." And they said, "You know, we were plant-eaters. We hardly ever ate grains, nuts, beans and seeds." They brought geologists in and they'd say, "Look at this. Fifty years, they ate nothing but fruits and plants, but now this ten-year period, they have indentations when droughts came." The geologist said, "Well, that's when there were droughts. So the plants dried up, the seeds were left and that's what we sustained ourselves on."'

According to Brian, 'There's no indication we were carnivores. And here's some of the facts. Six out of ten people on the planet now are primarily plant-based eaters. It's only 40 per cent of us that indulge, grossly overindulge, with animal-based food consumption. And if you look at the longest-living tribes of people, they eat either no animal-based foods or ritually, occasionally, do. Not because they need it. It's a ritual for them.'

Brian explained how he has watched the health benefits of people switching to a plant-based diet in his medical work. 'We've fed the sickest people in the world for sixty-five years and we've done clinical research to watch why the cancer goes away. Easy stuff like heart problems and diabetes, they go away so quick, your head spins. Not because we're some profoundly wise group of folks over here. We're just going back to the basics.'

As with Christiana's comments, for all the health benefits, Brian also saw how getting people to change their diet away from meat is difficult: 'Just yesterday, I was speaking to the group here, and I said, "What most of you miss is how this really pure life-style is really the most powerful way to heal. That you're looking for the stethoscope. You're looking for the big institutions. You're looking for the white coats. You're looking for volumes of science that makes no sense. But when I say to you, 'Think clean, eat clean and move well', somehow that's elusive to you. It goes in one ear and goes out the other."'

But despite this, Brian felt that through his work we were seeing a change in attitudes and lifestyles. 'I believe it's emerging. I believe that there's an ever-widening group of people on the global scale that are going back to the simple, the basics, the funda-mentals. It took from the nineties to now to understand that a plant-based diet is as important, if not more important than most things when it comes to the environment. For me', he concluded, 'the plant-based diet goes back to love. How can you really love when you're slaughtering an animal to eat it, thinking you need it?'

THE ASCENT IS ON

My ascent of Everest began by myself. It was early morning when I left Base Camp, one of those crisp, cold beginnings with the sky scored in blue and white above and the stretch of the journey and the climb laid out ahead in the white and grey of the snow, ice and rocks.

I'd learned on the acclimatisation climbs that I was slower than everyone else in the group. When I'd tried to walk at the same pace, I'd quickly worn myself out and increased my rate of exhaustion. It was important, I was told, to go at my own speed, and so that's what I set out to do. If I started earlier, then during the day, the others would catch up with me.

I didn't mind the early start. In fact, I really liked it. Back when I'd been training, I'd been getting up early to go to the gym, so in a way, this felt no different. I also really liked the silence and solitude. It gave me space to think, to remind myself what I was doing and why. And it gave me time, too, to record some messages for my colleagues and for my family. Even though I was away from my wife and children, making that recording felt as though it brought me closer to them. I knew the sacrifices they'd made to allow me to do this, and I reminded myself to make it up to them upon my return.

The climb out from Base Camp feels almost prehistoric. You're walking through these giant valleys that have been shaped and hewn by glaciers. With the Base Camp behind, all signs of modern life quickly disappear. On all sides there is potential danger. There's rocks and debris piled high from further up the mountain; as you're walking, there's a thud and a rumble and another pile collapses down. There are huge towers of snow that stick up from the groups like spikes – they're equally vulnerable to collapse. You're aware as you walk of this shifting landscape you're passing through. What once might have been a path is now a rockfall.

The easiest way to navigate through is to follow the yaks. The valleys are their home and they know all the best routes and cut-throughs. They're remarkable creatures, and it's only when you get up close that you realise how big they are, with the thickness of their shaggy coat and a pair of horns that twist up so the ends are pointing up towards the heavens. When a yak is on the move, you don't want to face them. Generally, they seem shy and tolerant creatures, but get in their way, and you're in trouble.

At one point, I was walking up a narrow trail, with rock to my left and a sharp fall down to the river below me. Ahead of me, I could see a yak walking towards me. The trail wasn't wide, and there was certainly not enough room for both of us. I thought the yak might stop, or turn back at the sight of me, but no – instead it dipped its head and carried on towards me. It wasn't charging me or anything like that, but it was making its movements clear. It was coming through, whether I got in its way or not.

I looked around. The climb above and the drop below left me little room for manoeuvre. I didn't want to retreat, but as the yak bore down on me, I didn't have much choice. Here I was, barely minutes into my ascent, and I was going back the way I came. As I retreated, I could hear the yak's breathing, its snorts as it continued down, getting louder as the creature got closer. In front of me, I could make out a possible ledge. I scrambled up, the loose rock giving way under my feet. I wasn't completely sure if the ledge was going to hold, but I clung on, hoping. Below me, with a snort and a strut, the yak continued down and past, not even giving me a second glance. As it passed, I slid back down onto the path, the yak's point well made as to who was really in charge.

● ● ●

The only people who had any sense of control over the yaks were the Sherpas. On the lower slopes between Base Camp and Advanced Base Camp, they used the yaks to carry the equipment.

The Sherpas are the unsung heroes of the mountaineering world. Put simply, without their support or assistance, there is no way that my attempt on Everest would have been remotely possible. Originally a nomadic people, the Sherpas settled from eastern Tibet in the Solukhumbu valley of Nepal in the 1400s, with Pangboche village being the oldest community. The name originally meant 'people from the east' and for many years they lived simply, isolated and cut off from the rest of the world. That began to change in 1907, when Sherpas were hired on an expedition for the first time. But it was the combined success of Tenzing Norgay and Edmund Hillary in 1953 that really cemented the Sherpas' reputation. And while the Sherpas gained international recognition, Hillary did much to promote the mountaineers' reputation with the Sherpas. He worked to improve the lives of the Sherpas, leading the Himalayan Trust from 1960 until his death in 2008, and which was responsible for building schools and hospitals in the region. He became known as the 'Sherpa King' for his efforts.

In 2017, a new scientific survey explored why the Sherpas are so good at climbing. This followed on from a previous study in 2010, revealing distinct genes among high-altitude living Tibetans, that were linked to oxygen metabolism and responding to a low-oxygen environment. This is a natural acclimatisation process coming from a line that has lived in the Himalayas for thousands of years. Over the centuries, this genetic line has developed ways of making the most of oxygen. For those of us who live at a more regular altitude, what happens when we climb is that a hormone called erythropoietin, or EPO, kickstarts the production of extra red blood cells, to help carry oxygen to our muscles. The 2017 survey showed that Sherpas, by contrast, increase their red blood cell count at a far lower rate. Not only this, but their mitochondria – the parts of the muscle that produce energy – were much more efficient than Westerners' mitochondria in converting oxygen into energy. At low altitudes, there was little difference between the metabolism of Sherpas and those living at lower

Yaks.

Sherpa team, Everest.

altitudes. But climb higher, and the genetic difference quickly began to kick in.

Today, the Sherpas are among the best climbers in the world. In 2011, Mingma Sherpa became the first Sherpa to climb all fourteen of the world's 8,000-metre peaks, and the first person to climb them all at the first attempt. In 2019, Kami Rita Sherpa became world record holder for the number of summits of Everest, reaching the peak for the twenty-third time. And while the window for climbing groups such as mine to scale Everest is a couple of weeks at best, the Sherpas are on Everest for the best part of three months, setting up tents and laying the ropes for the forthcoming expeditions in advance, and cleaning up the mountain long after the rest of us have gone home.

When we were staying in Kathmandu, we were fortunate enough to meet up with Karma Sherpa over dinner. He was someone who Lukas knew well and was responsible for helping to organise the Sherpas for our expedition. Karma was a fascinating guy, and had flown over from Japan, where he was living, to meet us. I asked him why the Sherpas were so special, and he explained that they were hardened from having to carry everything from village to village, including water.

At Base Camp, we met our group of Sherpas for the first time. It was quite the team. There were kitchen staff, and those responsible for the carrying of equipment. One of the cooks was particularly pleased when he worked out I was Turkish. His wife worked in Ankara as a chef for a family there, and he told me it was the first time he had ever met anyone Turkish on the mountain (I later found out there'd only ever been a handful of Turkish people who had been on the northern route before).

As well as the support staff, there were also the climbing Sherpas. Each of us were assigned two each, one to help clip us on and off the line, the other to carry the bulk of our equipment and oxygen (while our backpacks weighed around ten kilos, the Sherpas were carrying forty kilos in theirs, something I'm still not quite sure how they managed). Finally, there were the lead Sherpas, who worked with Lukas and the other guides, who managed the rest of the team.

In total, there were twenty-nine Sherpas on our team. As for our individual Sherpas, we didn't get to find out who they were going to be until we reached Advanced Base Camp. That process is a lottery – it's drawn out of a hat who is assigned to who. The reason for that is to stop the Sherpas from choosing their clients: for each of them, there's a $2,000 bonus if their climber makes it back down. So from Base Camp onwards, there was a bit of sussing out going on: I feel the Sherpas were sizing us up and discussing between them which climber they did and didn't want to get.

Before we made it to Advanced Base Camp, we were assigned two Sherpas at random. One of those assigned to me didn't speak any English, but the other was extremely chatty and talkative. He had seven children, he told me, and it was a struggle to feed them.

'Why do you keep on having them, then?' I asked.

'Because I don't think I'm going to live that long,' the Sherpa replied. 'And they're going to be Sherpas, so I don't think they're going to live that long either.' It was an answer said with a smile and a twinkle. When the final pairings were announced at Advanced Base Camp, I was disappointed that he wasn't one of mine. Instead, one of the Sherpas I got was another who couldn't speak English. And the other was the smallest, skinniest of the lot: how he was going to carry all the gear, I wasn't quite sure.

* * *

Advanced Base Camp sits at 21,300 feet, or 6,490 metres. From Base Camp it's a double leg of eleven and fourteen miles via the interim camp. It's a sweep up the valley, but the higher you get, so the amount of snow increases. So while at Base Camp, the bare rock gives you that lunar feel, Advanced Base Camp feels more polar. It's set out in the shadows of seracs: sharp packs of snow and ice with spiky peaks like a miniature mountain range. And while Base Camp is neatly divided into the tents of the different expedition teams, here the camp is stretched out in more of a long line. The comforts of Base Camp felt a long while ago; accommodation was now in tiny, two-man yellow tents, their thin canvas walls pulsing back and forth in the wind, which was beginning to pick up.

Advanced Base Camp was the end of the beginning, or the beginning of the end, depending on how you looked at it. This was as far as the yaks went with the equipment: from here on in, everything had to be carried by the team. I was relieved that I didn't feel anything like as bad as I'd done on my first climb up here. I was tired – all that scrabbling on loose stone and rock puts a lot of pressure on the legs and joints – but there were none of the warning signs of altitude sickness. The acclimatisation process, so far, had been a success.

About three or four miles out of Advanced Base Camp is where the real action begins. This is the point where the rock underfoot really gives way to snow and ice. There are a few benches set out here for everyone to change. This is where the crampons come out and for some people, the oxygen for the first time.

The crampons you attach to the bottom of your boots. If you didn't know what they were for, you'd think they were some sort of torture device: two racks of razor-sharp spikes to give you more grip on the ice and snow. I took my time putting them on – you have to get them on right, because if you don't and they become entangled, then they're almost impossible to fix. Once they're on, you have to get used to walking in them again. I'd practised a few times but even so, it takes some getting used to. You have to use this sort of duck walk to keep your feet apart; it's very easy to stab yourself by mistake with the spikes, and there are horror stories of people nicking a vein, cutting a main artery in their leg and bleeding out. On that first day of using them, I was extra cautious but still managed to stab myself several times – not as far through as my leg, thankfully, but my trousers were punctured by a rash of holes.

This is also where things started to get properly steep. From Base Camp to Advanced Base Camp you were climbing, but from Advanced Base Camp to North Col, it was as though everything got vertical.

It reminded me of the Wall in the TV series *Game of Thrones* – a solid, vertical structure at the top of the Kingdom of the North, designed to keep the Wildlings and White Walkers out. In the TV

Climb to North Col.

series the Wall is about 700 feet tall and made of ice, deliberately designed to be impossible to pass.

Looking at what we were facing to get to North Col, it didn't seem that different. I cricked my neck looking up to see what we had to climb. Against the wall of white was a thin line of black dots – other climbers zig-zagging their way up. At first glance I thought, how on earth am I going to get up there?

I clipped on to the rope. The fact that I had to do this told me how serious things were getting. From now on, if I fell, I needed that safety mechanism to stop me sliding down. Even with the crampons, it was slippy on the ice. With every footstep, you weren't completely sure if it was going to hold or give way. Behind me, my two Sherpas waited patiently. Every time I glanced back they gave me a nod or a thumbs-up of encouragement. But with their lack of English, I wasn't really sure what they were thinking about my progress. Were they wishing they'd drawn someone else in the lottery? Having them there felt reassuring, but it added pressure at the same time.

There was a lottery element to proceedings more generally. At times, with a crack and a thud, a serac would pull clear and crash down in a cloud of snow. Every time that happened, you held your breath and thanked the stars that you hadn't been standing under it. Anything dropped from above could be similarly deadly. At one point, a small pocket camera plunged past me: it couldn't have been more than thirty centimetres away from my face. I looked up to see who had let go of it, but they were so high above, I had no idea.

Fred and Marco were consistently ahead of me – a good hundred metres or so. When they stopped to have a rest, I'd catch up. But with the possibility of seracs crashing down, you didn't really want to hang about any more than you needed in order to refuel and catch your breath. At times I was grateful for the footsteps left in the snow: they gave me a trail to follow. But everyone's stride was different, and again I was reminded of the importance of going at my own pace. It was easy to stretch too far and leave yourself exhausted.

I was beginning to really feel the altitude by now. Without the oxygen, I could feel the shallowness of my breath and that little bit of light-headedness. I was feeling tired too – the lack of oxygen meant less energy, less strength in the muscles and a natural slowing down. Ahead of me, I saw Fred taking oxygen. He'd said beforehand he was going to get to 8,000 metres before he'd succumb; we weren't at 7,000 metres yet and he was making a start. If it was good enough for him, then it was for me too. I noticed some of the Sherpas, too, were getting their masks out. I followed suit, the burst of energy coming at exactly the right moment.

In the sunshine, the colours seemed to get brighter: the brilliant white of the snow, the deep blue of the sky, the reds and yellows of the mountaineering gear. The climb up was relentless; there were many moments when I was happy to be attached to the rope, either because of my feet giving way or so I could use it to grab on to and pull myself up and over a ridge. Occasionally I looked back to take in the view, and each time I regretted it. It wasn't that it wasn't stunning, but the lurch of vertigo, the realisation of how steep and long the drop was, only served to shake my nerves still further.

I thought it would never end. It took eight hours of tortuous, painful climbing to get to North Col. This camp is up on a ridge and much more exposed. Compared to Base Camp and even Advanced Base Camp it felt tiny, a small smattering of yellow tents clinging on to the mountainside for dear life. North Col sat at 7,000 metres. I was higher than on the summit of Aconcagua, but getting to the top of the Argentinian mountain was nothing in comparison to this.

The harsh realities of mountain climbing kicked it. There were no facilities there; the only way to wash was to wipe yourself down with wet wipes. That was unpleasant enough, but I was so exhausted from the climb that even basic actions felt like a huge effort. The danger to life was everywhere. Even when you were sitting down, you were still hooked into the line, in case something fell from above or a gust of wind took you over. The only person not hooked in was Mingma Sherpa, who strode around the camp with a reassuring confidence.

Fred, Marco and I, by contrast, sat there, frozen in our thoughts in both senses. At least we had hot food here to eat: lentils, rice, yak meat. It was filling and hugely welcome; I could feel the warmth spread through me as I ate. You're meant to eat as much as you can, packing yourself full of energy, but I found the second plateful was hard to swallow. I needed to eat, though: much of the food I'd brought with me to keep me going was not fit to eat. I'd had these custom-made bars put together for me by a close friend of mine. They were a mixture of macadamia nuts, honey, maca powder and dried fruits. The bars were packed with superfoods, were super healthy and tasted great. Unfortunately, they had frozen solid. They might be useful if you needed a hammer, but were completely inedible.

Lukas came down to sit with us, along with another guide. He gave us a weather report, which was still looking OK, and he pointed out the route we'd be taking next. From North Col, the route was along the ridge to the higher camps, and then the final Three Steps to the summit. From further down the slope, you didn't really get a sense of how steep it was, or how sheer the drops were on either side. Up closer, it looked terrifying. The wall had been difficult enough, but somehow, this was harder again still.

'It's time for the talk,' Lukas said. That sounded ominous, and it was.

'From here on in,' Lukas continued, 'you're on your own. You have your Sherpas with you, and they will help you and look after you, but if anything happens to you, we can't get you down. You can't get a helicopter up here, and you've seen what you've just climbed up: no one is going to carry you down from here. So if you have any problems, you break your ankle, have fatigue, exhaustion, that's your problem to deal with.'

I double-blinked at all of this. I'd thought that signing up with Lukas, with the best team available, that if anything went wrong they'd be there to help look after me. But even with the best organisation and equipment, I realised that help could only go so far. Lukas was making it clear that to go on further from here was our decision, and our risk to take.

After Lukas had left us, we continued to sit there in silence. I didn't feel hungry any more. Lukas's talk was the first time that the danger of what I signed up for really hit home. All day I'd been wondering if I was going to make it up. Now, with a shiver, I was terrified about whether I was going to make it back down.

A NATURAL SOLUTION

'You might assume that the challenge for Shinkansen is how to make it run faster … in fact, the greater challenge for us is how to make it run more quietly.'

Back in the mid-1990s, Eiji Nakatsu was in charge of the technical development department of Japan's Shinkansen, or bullet trains as they are more commonly known outside of Japan. Over the previous three decades, the country had led the way in terms of high-speed trains. The first bullet trains were launched in October 1964, just in time for Tokyo's hosting of that year's Olympic Games. Within a year, they had halved the time it took to travel between Tokyo and Osaka from six hours forty to three hours ten. By 1967, the high-speed trains had been used by 100 million passengers; a decade later, they reached the 1 billion passenger mark.

But by the 1990s, the Shinkansen had a problem. The trains had the capacity to go faster, but they were limited by the amount of noise that they were making as a result. The government guidelines required a maximum level of 70 dBA (decibels, A-weighted in the industry jargon). One particular issue was the sonic boom sensation when the bullet train entered a tunnel. What was happening was that when a train entered a tunnel at speed, it pushed the air in that had built up in front of it. The compressed air created a sound wave, and as this was released at the other end of the tunnel, it created a boom noise – ironically, given the train's nickname, it sounded like a loud gunshot. At the same time, the air resistance the train encountered from entering the tunnel slowed it down; it was like 'wading into water', according to one account. On top of this, given Japan's mountainous terrain, the train lines often involved a lot of tunnels. The Osaka to Hakata line that Eiji Nataksi was hoping to improve was tunnelled for around half of its length.

'One day', Nakatsu remembered in a 2005 interview with Japan for Sustainability, 'I happened to see a notice for a lecture in a newspaper. I had the honour of meeting Mr Seiichi Yajima, then an aircraft design engineer and also a member of the Wild Bird Society for Japan. From him I learned how much of the current aircraft technology has been based on studies of the functions and structures of birds.'

For Nakatsu, a keen birdwatcher, this revelation was a light bulb moment: 'I was struck by the amazing functions that have been developed by living things … from then on, "learning from nature" became a recurrent theme for me.' He tried to apply this theory to the bullet train problem: 'The question occurred to me – is there some living thing that manages sudden changes in air resistance as a part of daily life? Yes, there is. The kingfisher.'

The kingfisher is one of nature's most beautiful birds, as anyone lucky enough to have seen its flash of blue near a river-bank can attest. But it is also a miracle of design, able to dive at high speed from air into water (800 times denser) without losing speed and hardly making a splash. When Nakatsu and his team studied why this was, they worked out that it was due to the shape of its beak. Pressed together, the two beaks form a sort of squashed diamond shape, the length and narrowness of its cone allowing it to reduce the impact on hitting the water, maintaining its speed and helping catch its fish. As Nakatsu's team experimented with reshaping the train's nose, the design that copied the kingfisher's beak was by far the most successful design. It allowed the train to travel more quietly and faster, and use 15 per cent less energy than before. In March 1997, the redesigned train made its commercial debut, running at a then world record speed of 300 kilometres an hour.

'I came to a clear realisation that answers or clues can be found in nature,' Nakatsu reflected in 2005. 'I studied engineering at school and worked as an engineer, but when I look back at my school days I wish I had studied biology more. I would urge schools and universities to teach students, especially engineering students, more about the survival strategies of living things as expressed in their shapes and functions.' He goes on to quote a book recommended to him by Seiichi Yajima: "'A tree, a blade of grass, a bird or a fish, all can be brilliant and everlasting teachers." This philosophy continues to inspire my work as an engineer.'

• • •

In this chapter I want to explore some of the different ways that nature can be utilised in the fight to combat the climate crisis. Because, for all the potential and the role technology plays, as with the case of the kingfisher and the bullet train, sometimes looking at the natural world for inspiration can be just the starting point we need. And even such humble species as trees, grasses and algae can play their part in saving the planet.

Such ideas are known as nature-based solutions, or NbS in the jargon. There is an ongoing discussion as to precisely what constitutes a nature-based solution, but broadly, they involve increasing the amount of biological material that can remove CO_2 from the atmosphere – so planting trees, converting to no-till farming methods, or an architect designing a roof garden that would otherwise be concrete are just some examples. According to a recent United Nations report, nature-based solutions have the potential to remove a third of excess carbon released into the atmosphere each year.

Tom Chi's remarkable career has taken him from Silicon Valley to leading some of the most innovative environmental schemes on the planet. Educated in electrical engineering, he worked for Microsoft on Outlook, as a Senior Director for Yahoo!, then as Head of Experience of Google X, working on products such as Google Glass and self-driving cars. Since then, he has moved over to social/environmental entrepreneurship, through a mixture of direct innovation, investing, mentoring and coaching. Spend half an hour in his company, and you realise immediately that here is someone both fizzing with ideas and passionate about saving the planet. You can see immediately why he gave himself the job title of 'Lab Director of the Planet'. He has since founded a venture capital firm, whose goal is to invest toward a world where 'Humanity is a Net Positive to Nature.'

When I spent some time with Tom talking through some of the projects he is currently working on, it was clear that like the experiences of Eiji Nakatsu, the more he developed ideas to deal with climate change, the more in awe he was, as an engineer, at just what nature was able to achieve.

Take the technical challenge of finding a way to tackle carbon dioxide through the use of air capture machines, for example: 'If you want to have something able to capture a lot of carbon in a short amount of time, you have to consider the collection area and how much carbon dioxide mass is coming through it. I refer to this as mass flux. If you wanted to achieve effectively large mass flux for civilisation using machinery, you'd have to build a huge number of facilities. Even the largest industrial facilities that we have in the world might be one square mile or half a square mile. That's the largest we can do in terms of the facility itself. This means the aperture we have to collect carbon dioxide is tiny compared to the surface areas of the earth, so your only chance of getting enough mass flux through this small mechanical aperture is to actively force a ton of air through it. This uses a lot of energy and makes it pretty expensive to do.'

'Or you can have a huge aperture and collect at a more modest rate. That,' Tom explained with a smile, 'is what biology does. A forest has got a huge collection aperture (every leaf, plant, living thing). Because of this, you can achieve a way bigger mass flux in way less time if you went on the biological route over any type of mechanical building. And that is apart from the mining damage that comes from all the steel and aluminum, the energy it takes to run the fans and release the carbon from the sorbent, and all the other logistics and operating costs for these industrial facilities.'

Trees play the central role in one of Tom's plans to reduce carbon emissions. His goal is both simple and audacious at the same time: to plant a trillion trees to help soak up the excess carbon that we have created. Tom estimates that over the course of human history, we've cut down somewhere between 2 to 2.5 trillion trees. So planting 1 trillion is returning 40 per cent of that total. According to Tom's figures, humans have emitted 2.2 trillion tonnes of CO_2 since 1750: 26 per cent of this has been absorbed by the oceans and 29 per cent by the land. That leaves

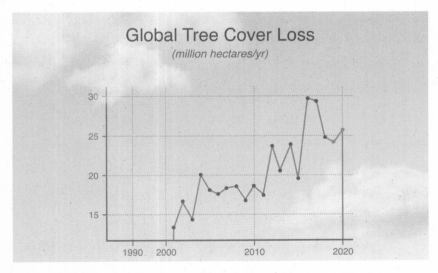

'World Scientists' Warning of a Climate Emergency', *BioScience*.

(Source: William J. Ripple, Christopher Wolf, Thomas M. Newsome, Phoebe Barnard, William R. Moomaw)

around 45 per cent remaining in the atmosphere. The mass planting of trees is a way of paying off this one trillion tonne carbon 'debt', as Tom describes it.

So how do you go about planting a trillion trees? The short answer is that you plant 20 billion a year for fifty years. How you go about doing that is with the use of drones. Here's where Tom's understanding of technology comes in. He is involved in a company that has developed seed-firing drones that can plant 120 trees a minute. The seeds are fired out from the drone in a biodegradable pod. Using digital intelligence, the drone can be directed to drop the seeds in the right location. And because they are flying, they are able to reach spots that a human planter might not otherwise be able to get to. Using twenty-five drones, Tom estimates that a million trees could be planted in just five hours. With an army of 9,000 drones (requiring a team of about 450 to fly them), the 20 billion a year target could be hit working ~200 days a year.

It sounds like a hi-tech solution, but as Tom says, 'The most important technology in this system are the seeds. The organisms that we're putting back are the most important technology. Compared to the complexity of a seed, creating drone planting implements and developing scanning, monitoring and maintenance mechanisms, it's all easier work than what nature has already contributed.'

• • •

For Mark Herrema, the genesis of his carbon-saving idea came to him when he was still at college. 'I got really sick when I was in my junior year at Princeton. I started losing a lot of weight. I had a bunch of really bad symptoms. I was told I had some internal bleeding. I was seeing tons of different doctors, and nobody could figure out what was going on. In the end, it turned out that I had coeliac disease.

'The summer between my junior and senior year, I was doing a ton of research to try and figure out what was going on. One of the really bad symptoms of coeliac is really bad indigestion, and in my research, I came across a newspaper article about methane emissions from cows. I looked at this article, and there was something really striking. It talked about how much each cow burps. It's about 600 litres of methane per cow per day.

'The reason why that mattered so much is we talk about climate change, and carbon emissions in quite general terms a lot. But this number, it was so specific that you could do math with it. I worked it out to about $20 per cow per year emitted. If you have a 1,000-cow farm, that's $20,000 of value going into the air. I remember calling up my friend Kenton Kimmel, and I said, "Hey, you know, it seems to me that everybody's talking about taxing carbon or trying to pump it underground. But what if we could turn carbon that was going to go into the air into useful products?"'

It's a simple but brilliant idea: the challenge, of course, was how to go about doing it. Mark and Kenton's discovery was to use a micro-organism that is found in the ocean, and which consumes methane and carbon dioxide as a food source. As the micro-organisms feed on the greenhouse gases and grow they produce a molecule within their cells called PHB. If you can extract the PHB and purify it, the substance is meltable: a sort of naturally made material that can replace plastic, which Mark and Kenton call AirCarbon resin. Because the AirCarbon is naturally rather than synthetically made, it breaks back down completely in the environment as a food source for micro-organisms, similar to a tree leaf. And because it is being produced from greenhouse gases, the whole thing is a carbon negative process.

Achieving all of this was easier said than done. 'It seemed easy on paper,' Mark remembered. 'It took us about ten years to work out how to do that. We started with a test tube and a crazy idea. We moved into a university lab for a couple of years. Then we raised a small amount of capital, built a pilot production facility. We spent what felt like a lifetime in that facility just trying to figure out how to make this process efficient.' All that hard work and investment has paid off. Today, Mark and Kenton's company, Newlight, has a two-and-a-half-acre operating facility employing about eighty people. 'We have eleven conversion reactors, eight lab units, two pilot units and our largest fully integrated commercial-scale production line.'

The magic micro-organisms are ones that Mark and Kenton found through 'a lot of field tripping. We looked around at waste water treatment plants, and estuaries and oceans. Back then, we didn't have any money to do any sort of exotic field tripping. So our bugs are southern Californian bugs.' The company is experimenting with a number of different items, but have started by making two products whose plastic equivalents have blighted our oceans: straws and cutlery.

Accountability is important for Newlight. They got their air carbon material independently verified as carbon negative by both the Carbon Trust and SCS Global. They also got official clearance to call it ocean-degradable: this means that a product dissolves as fast as paper in ocean water. They've also developed a set-up using IBM blockchain tracking for their fashion products, so that consumers can track back through the process of manufacture, and feel confident of the product's environmental credentials.

It's amazing what Mark and Kenton have already achieved, and clearly the potential to replace synthetic plastics with a natural alternative is huge. And like Tom Chi's idea for the planting of a trillion trees, it comes about through a combination of the latest technology and a key natural component. The micro-organism at the centre of the project may not be much to look at, but it is very much the star of the show.

• • •

Another unsung hero in the battle against climate change is something else that lives in the sea, and is as easily overlooked as Mark's micro-organisms: seagrass. *Posidonia oceanica*, to give it its Latin name, is one of the oldest living organisms on the planet. In 2012, DNA analysis on seagrass living on the Mediterranean floor near the Spanish island of Formentera revealed it was 200,000 years old. It's a slow-grower: seagrass can take up to 600 years to cover just 100 metres of ocean floor. But it has never been in any particular hurry to do so; without competition or predators, it has been able to grow happily in its underwater meadows. That is, until recently. A mixture of rising ocean temperatures, trawler damage and coastal development had depleted its stocks, which are down 10 per cent in the Mediterranean, and more than 90 per cent in UK coastal waters.

When I spoke to Zafer Kizilkaya, head of the Mediterranean Conservation Society, he told me about their part in a joint project to chart the amount of seagrass in the sea. Employing a satellite that has been recently launched into orbit by the European Space Agency, the project is using the satellite's increased optic quality to measure and map the seagrass meadows from space. 'Mediterranean seagrass has amazing properties in carbon sequestration,' Zafer explained. 'The seagrass has extraordinary capacity to store carbon dioxide, 20 tonnes per hectare. That's five times more than the next plant, which are mangroves.' Compared to tropical rainforests, seagrass stores carbon an amazing thirty-five times faster. And while seagrass covers just 0.2 per cent of the ocean floor, it accounts for 10 per cent of the ocean's carbon storage.

'The most amazing part of it,' Zafer continued, 'is that they're not storing carbon in their leaves, they're not storing the carbon in their roots, they're storing the carbon just in the sediments. So it never ever comes back. If you cut down a tree, you burn it or it becomes the cane, the carbon gets back to the atmosphere. But the seagrass deposits in the sediment, and it never comes back to the atmosphere. It is very important to save them.'

And not only does seagrass help in storing carbon, a recent study by marine biologists from the University of Barcelona revealed that they play their part in tackling plastic pollution in the oceans as well. Seagrass appears to trap plastic pollution in small bundles of fibres called 'Neptune balls.' According to reports, seagrasses collect 900 million plastic items a year in the Mediterranean alone.

As well as environmentalists like Zafer working to protect seagrass at current levels, others are exploring the possibility of creating new underwater meadows. The Novograss Project is one such trial off the coast of Denmark. A similar scheme is underway in Wales, supported by WWF and Sky Ocean Rescue, to create a 20,000 square metre meadow. If these projects are successful, then they have the potential to be rolled out more widely.

• • •

Hearing about these various experiments around the role that different species can play in combatting climate change led me to wonder whether we should be thinking a little more widely about nature. Is there something in our relationship with the natural world that has gone astray over the years – and if so, how do we go about rebuilding this?

To answer this question, I turned to 'three wise men' who over the years have helped me at different points in my life. Patrick Creelman is the founder of Pure Yoga, the largest yoga franchise in the world, and a bit of a celebrity in spirituality and yoga terms. Alexandre Tannous is an extraordinary thinker – a specialist in sounds and music and a generalist in life and philosophy. And Nando Vila is a Colombian healer and environmentalist, whose retreats I have been fortunate enough to participate in. And although I spoke to each of them separately in my research for this book, their answers about our relationship with nature told a similar story.

Patrick described how nature was firmly embedded in his teachings and practices. 'In my life there are two parallel paths. One is the Buddhist path, primarily the Bhagdryana from Tibet, and the other is a Hindu path, from India. In the Hindu path, they break down the cosmos into two main concepts: one is spirit, *purusha*, and the other is *prakriti*, which is nature. Under this model of yoga and spirituality, nature is earth, water, fire, air and ether.

'What that means is that our truest nature is our spirit, the light of our consciousness. The covering of our spirit is our body, our mind, our heart. And if you extend that further, then it's the earth around us: nature extrapolates and differentiates further and further to the whole cosmos. But the centre of that for each human is the light of our own awareness. So that's our spirit, and then the rest is nature.'

Patrick explained how building this relationship with nature is important, and the differences you noticed when it happened. 'Without nature, the body dies. It's done. We need all the elements. And interestingly, when you start taking a keen interest in your body, almost 99–100 per cent of the time, what happens is people care about the extension of nature. You can't really care about this as a vessel for the light of your soul if you don't care about the earth around you. Take the natural health boom that's happening right now with keto, paleo and all that. People want the right food. And from that can come appreciation.'

Patrick argued that 'yoga and even the modern cognitive sciences and neurosciences are an amazing link to the natural world … When we tap into the greater cycles of the world, the climate, the seasons, it's not just tuning into nature, it's tuning into the point that humanity and nature are so clearly one. They come together. It's a package deal? For Patrick, it is very easy to lose this relationship with nature. But by focusing within, and on our spirituality, it is possible to build it back up. And when we do, the strength within ourselves leads on to a stronger interest in the planet around us. For all the hi-tech solutions to the climate crisis, a simple shift of thinking and greater spirituality can also help to change our behaviour. This would allow us to understand and appreciate nature more and maybe, like Eiji Nakatsu, find the inspiration we need.

'I'm from Colombia,' Nando Vila told me about his background, 'from the middle of the India part of Colombia. We have a tropical rainforest there. This was our way: instead of going to the city, to bars and everything, we would retreat over there and be in nature. I think the connection with nature has been the teacher. Now, Colombia has more than a hundred tribes. I got to learn from some of them, from the Amazon, but also the Kogis, the Arawakos. I learned some of their ways of healing. I learned that keeping yourself in connection is having the right tools to connect to nature.'

Like Patrick, Nando described how he felt we have lost this relationship with nature. 'We are in a very unique moment of a planetary life that we decide the future of. So we all have to become experts on climate change. But we are all experts on nature because nature is in our cells. This adjustment period of adaptation at this time demands waking up to what we are;

as a species here on Mother Earth – to remember all the evolution in time that took to come to what we are, all the changes and transformations, all the shapes shifting and states of forms, textures, colours and the infinite diversity of nature. We might not remember it because these memories have not been acted on, as our contact with nature is continually separated and distanced from our daily lives.' I asked Nando when this connection became ruptured. He mentioned the start of the Industrial Age as being important, but went back further still: 'We can even trace it to 2,000 years ago when we decided to make cities based on our agriculture promise of providing supply and health. From then, we don't have to really worry about the fear of not having food. That created a very comfortable sleeping spirit time where we don't have to put our hands on the earth.'

So how do we go about reconnecting with nature? 'I think we need to re-educate ourselves. Children, when they come with a free spirit, they come naturally with that connection. But we come to a programme of education where nature is separated from us, where we learn to read numbers, but not learn to read nature, the clouds, the animals, the trees, the seasons, the little animals that come, they all have a message there. We don't read that anymore. You might think you are very healthy and connected in nature; but if you don't go to the river, if you don't swim in the ocean, if you don't go to connect with the plants, and see all these from your heart … you're not feeling it.'

Alexandre Tannous also echoed these thoughts again in his own inimitable way. 'We need to bring science back to natural philosophy', he explained to me. 'Natural philosophy is the place where science came from, but we lost connection to nature and spirituality … one way to wake up and understand what we need to do to reconnect to the universe is to seek inner guidance under someone who works with integrity. What we need to do is invest. We need to create our own form of shamanism, a shamanism informed by science, by academic field, by philosophy, by neuro-science, by Eastern philosophy, Gnosticism, Hermeticism ….' In a similar way to Nando, Alexandre describes how this disconnec-tion with nature can stretch back over time ('Since Babylon') but has accelerated over the past 200 years or so. Even so, hearing him speak, he was positive for the future: 'People are fed up, people are waking up', he said.

There was a lot to take in from these conversations with Nando, Patrick and Alexandre, and I feel I can only scratch the surface here. But there was something in what each of them was saying that felt worthy of consideration. And though exploring different traditions and philosophies might not feel a conven-tional way to tackle climate change, if their ideas help to bring us back closer to nature, that can only be a positive thing. Or as Alexandre summed it up, 'Energy flows where intention and attention go.'

Their thoughts reminded me, too, of the words of Patriarch Bartholomew, the leader of the Eastern Orthodox Church, which has 260 million members globally. Known as the 'Green Patriarch', Bartholomew has spoken often about the importance of protect-ing the planet, describing it as a moral duty. Addressing the

Young Presidents' Organization in 2019, Bartholomew stressed how looking after the planet went hand in hand with how we treat each other. 'The way we respond to the natural world is intimately connected to the way we treat human beings,' he argued, adding, 'the willingness to exploit the environment goes hand in hand with the willingness to ignore human suffering.'

Bartholomew challenged those at the conference, and all of us more widely, to step up. 'Do we honestly do all that we can to leave as light a footprint as possible on this planet for the sake of those who share it with us and for the sake of future generations? Today there are no excuses for our lack of involvement. We have access to detailed and instantaneous information; and the alarming statistics are readily available. So if we do not choose to care, then we are not simply indifferent onlookers; we are in fact active aggressors.'

THE DEATH ZONE

The climb out from North Col was a bit like walking up a ski ramp. The snow was thick and I could hear the crunch of the crampons underfoot and feel the stretch on my hamstrings as I pulled myself up. Around us the clouds swirled on and off the slopes, one moment shrouding the mountain from view, the next revealing its jagged edges of grey and white against the rich blue of the Himalayan sky. I wasn't sure which was more frightening: looking up to see what was still to come, or looking down, to take in what would happen if I fell.

At times, it was an isolating haul up. Fred, Marco, Gabriel and I were walking at our own pace, and pretty soon we were separated as we made our way up. I was with my Sherpas, but with their lack of English, conversation was limited. The fact that by this point everyone was masked up and breathing in oxygen reduced interaction still further.

Inside the mask, all you could hear was the sound of your breathing. Outside, apart from the crunch of snow, it was the wind you heard above all. This would rise and fall, one moment calm, the next buffeting you and making you grateful for being clipped to a line. When it rose, the way the wind was funnelled meant that it made a strange, siren sound – it was deep and whistling, quite other-worldly and unnatural. You'd hear that noise and then, up above, you'd watch as a plume of snow was pushed off the summit. Where is all that snow coming from? I kept asking myself. It felt like an endless battle of the elements – snow at night, wind during the day.

Above me, I could see the thin trail of the rope and the black dots of the climbers ahead, like ants. At times the rope zig-zagged to take the sting out of the slope, but even that made little dent in what we were climbing up. Although the climbers were spread out, it was still noticeable how many of them there were. The numbers were nothing compared to the route up the south side, but even so, they were both considerable and a hazard. With everyone following the same route up, you could only go as fast as the person in front of you. To get past someone, you had to take the risk of unclipping yourself to work your way round. My bigger problem was with people behind me, hurrying to get past. That put pressure on me to go faster, ignoring everything I'd been told about going at my natural pace. Some climbers were more impatient than others, and not without reason: getting stuck in a queue and not moving was dangerous, if not deadly. So there was little time for niceties; it was like a golf group coming through, except there was no waiting around for an invitation to do so.

The crowds on the mountain came down to commercialisation. It was down to Nepal and China as to precisely how many passes to hand out. These weren't cheap to buy – indeed, they made up a good percentage of my overall fee. The more passes the authorities issued, the more money they made. But the more passes they handed out, the greater the risk for overcrowding, especially when the weather window was as narrow as it was in 2019. It was all a far cry from the empty climbs of George Mallory a hundred years before.

All this commercialisation inevitably had an environmental impact, too, which our party at least tried to allay. We began by offsetting the carbon footprint of the expedition through the purchase of carbon credits. Fred, meanwhile, launched a campaign on LinkedIn, donating €1 for every like he got to Marion Chaygneaud-Dupuy's Clean Everest project. He raised €20,000. As for our own waste on the mountain, the expedition made sure that everything was collected and brought back down, leaving the mountain as unspoilt for others as we could.

The good news was that I was feeling stronger than I expected. Whereas on the lower slopes, the marathon-running experience of Fred and Marco saw them pulling away from me with ease, now what was required was more a balance of endurance and strength. All that work in the gym I'd been doing since early the previous summer was beginning to pay off. Which wasn't to say that the day wasn't exhausting – by the end of it, I was absolutely shattered – but I found I was matching my friends' pace more closely now. Rather than disappearing as dots in the distance, they were more in range, struggling just as much as I was.

Climb to Camp II.

The route from North Col to Camp II is a full day's slog: you start at seven in the morning and reach camp at four in the afternoon. As the day progressed, so the terrain shifted. The higher you got, and the steeper the slope, the more the snow thinned out. It was simply too severe to hold. That made it increasingly difficult to walk. While you were moving through the snow, there was plenty for the crampons to grip on to. Once that had gone and you were onto rock, everything became a bit more slippy. Having cursed wading through the snow earlier in the day, now I found myself wishing it was back.

If I thought North Col had been scratchy compared to Base Camp and Advanced Base Camp below, Camp II was something else. It wasn't really a campsite at all, more a scattered selection of tents stuck at 45 degrees on the side of the mountain. And while North Col was relatively sheltered and cocooned by the snow and corniches around, Camp II was much more exposed.

We were high now. It was disappointing, in a way, that for a full day's assault, we had only climbed another 500 vertical metres. But that showed you the severity of what we were now traversing. Ahead of me, as the clouds skimmed on and off the summit, I could grasp a sense of the final route to the peak. My eye followed the line of the ridge and, with a gulp, I recognised and followed the Three Steps to the top. From here, it looked impossible.

But for now, I just wanted to rest. Camp II was going to be the last night's sleep before we pushed on all the way to the summit. The next day would see a slog up to Camp III, a pause for a few hours from late afternoon to early evening, and then the overnight journey to the top, reaching the peak the following morning. Then it was another full day's descent. Above Camp II, the altitude tipped over 8,000 metres – the so-called death zone, where human life can only survive for a limited amount of time. I was glad we were sleeping below that, and have the knowledge that I was safe for one more night at least.

By the time I got to Camp II, I could barely move. I was physically on the edge of collapse. I shouted to one of the team Sherpas, asking which one my tent was. I was pointed to one sat up at an angle and flopped down in front of it. Then one of the other Sherpas appeared, and pointed further up the line.

'Sorry Hakan, that's your tent up there.'

I shook my head. 'I'm not moving,' I said. 'I'm staying here.'

The Sherpa tried to argue, but from the look on my face, quickly gave up. I crawled in and sorted myself out. I blew up the mattress, got my sleeping bag out, and then used my small bag of spare clothes as a pillow. It still wasn't comfy, but I didn't care. I was exhausted. I took my crampons off and clambered in to try and get comfortable. Outside, the wind was whistling again, that eerie siren sound. The canvas walls of the tent pulsed in and out. That was the last thing I remembered as I drifted off.

●   ●   ●

'Hakan! Hakan! Are you OK? Can you hear me? Say something.'

I was out cold when I felt a shockwave thump through the tent. A deep boom, then a crash and a whooshing sound. I was so tired and in such a deep sleep, I was confused as to what was going on. I could feel the wind on my face, a drop in temperature, and then shouts. Lots of shouts.

I struggled to blink myself awake. That only confused me further. Where was I? I didn't seem to be in the tent any more, but instead could see the darkening of the sky above. Then came the faces: Sherpas looking terrified, and Gabriel peering over me like he'd seen a ghost.

'Hey,' I said drowsily, trying to shake myself to. 'What? Huh?'

'Can you feel your legs?' Gabriel turned around to one of the Sherpas. 'He's speaking. He's alive.'

'Hey,' I said. 'Hang on. What's going on? What's happened to …'

It was now I took in the surroundings. The tent I'd been sleeping in was no longer there, but ripped in two, its flimsy canvas sides flapping in the wind like yellow flags. The ground seemed strangely harder and I glanced down to my blow-up mattress, except it was no longer there: there was a hole in the side, where it had been reduced to rags. More familiar scraps of material fluttered past, and I realised it was the bag I'd been using as a pillow. I was wide awake now.

'Gabriel?' I asked. 'What's happened?'

'This,' Gabriel said, leaning over and pointing to a huge rock that was now sitting in the middle of what had been the tent.

'My God,' I said. 'Where did that come from?' I sat up and tried to move it. The slab of granite was heavy – ten, fifteen kilos at least.

Gabriel and I both turned to look up the mountain. Because I'd been so tired and fallen into the first tent that I could, the one I'd chosen was quite close to the climbing line. Someone

Camp 2, Everest.

further up the mountain must have dislodged it, and it had come crashing down. Looking at where the rips in the pillow and mattress were, I realised how close the rock had been to hitting me. Another inch or two, one roll across in my sleep, and, well, it would have …

'You're lucky to be alive,' Gabriel said.

I gulped. Even if the impact of the rock didn't kill me, the lack of medical equipment would have finished me off. There were no supplies that high to deal with a serious head injury, and no way of whisking me off to hospital. The Sherpas, I'd discovered, didn't even carry a proper first aid kit with them. They had adrenaline pens and drugs to deal with mountain sickness, but that was pretty much it. For head injuries, there was no serious kit to help.

As the enormity of what had happened kicked in, I definitely didn't need an adrenaline pen. My heart thumped from delayed shock. Gabriel and the Sherpas helped me to gather my things and I was moved to one of the Sherpa tents, further away from the climbing line. They shuffled round to bunch up in one of the bigger tents.

But even as I was catching my breath from having so narrowly escaped death, the drama wasn't over. As we were carrying the stuff up, Gabriel called over to me. He was shouting and pointing at another tent, where I could see two climbers were collapsed outside.

'Hakan! Give me a hand helping them up!'

I rushed over to help him prop the climbers up against the rock. Their bodies were heavy and limp and as we leant them back, I gasped. Their eyes had rolled back so all you could see was the whites of their eyes. They were alive, I could hear that from their groaning, but they were not in a good way at all.

'They're running out of oxygen.' Gabriel looked at their gauge to check the levels. 'We need to get them some more.'

I shouted up to our Sherpa who was nearby. He came rushing down to join us.

'Where are your Sherpas?' I asked the two climbers. They could barely speak.

The Sherpa unhooked his radio and shouted instructions into it. I wasn't sure if he was speaking in Sherpa or Nepali. A voice crackled back.

'What are they saying?' I said, when the conversation finished.

'They're ahead,' the Sherpa said. 'They're coming back.'

'How far away are they?' I asked.

The Sherpa shrugged. 'They'll get here as quickly as they can. In the meantime, do not – do NOT – give them your spare oxygen.'

It was brutal. As the minutes ticked by, Gabriel was in tears. The climbers were delirious. They were Americans, that much I could gather from what they could whisper and groan. They didn't appear rookies like me, but big strong individuals, with all the right gear. And yet, here they were, without their Sherpas. Ours stayed with us, barking into his radio at intervals. I think he was there almost to watch us as much as he was keeping an eye on them. For him, who made his living on the mountain, the choice was straightforward: it was between the possibility of these guys dying now and us dying later. For Gabriel and myself, the choice was more heart-wrenching.

After twenty long minutes, the Americans' Sherpas finally arrived back in camp. As with our Sherpa, there was little emotion there: just a matter-of-fact acknowledgement of the situation. They fixed up the climbers as best they could, and I watched with relief as they finally got some fresh oxygen on board. They were OK – for now. I guiltily felt relief that I'd gone with Lukas: oxygen and organisation were everything, and our expedition was doing everything they could to keep their climbers safe.

By this point I was both exhausted and scared. To be so close to death myself, and then to see others in the same situation, it hit home just how dangerous the situation I'd got myself into was. The words of the Aconcagua guide floated through my mind again. *Everest is something else. Seriously, Hakan, you have no idea. If you try and climb Everest, you'll die.* I thought of Stephanie and my children. What on earth was I thinking? How am I going to get out of this? Am I going to make it back down? Am I going to see my loved ones again?

I wanted to sleep, but the adrenaline running through me was putting paid to that. The aftershock of coming so close to dying was reverberating through me. I'd no idea who dislodged that rock, and I don't suppose they even realised they'd done it. I knew this was my last chance to get some rest, but that only ramped the pressure up further. Every blast of wind or rattle of scraping of rock had me sitting up bolt upright, terrified that another slab of stone was going to come crashing through and finish me off.

• • •

If I had any doubt that the climb from Camp II to Camp III took me over 8,000 metres and into the death zone, the sight of bodies lying by the side of the path was a grim reminder.

Around 300 climbers have lost their lives on Everest in total, with the vast majority on its upper slopes. And while improvements in technology and techniques have in many ways made the climb safer, that has been counterbalanced by both the increasing volume of climbers attempting to summit and the dangers these additional numbers create. Rather than going down, the average number of fatalities has crept up in recent years.

There's no sign when you pass the 8,000 metres mark, but once you're above that height, the clock is ticking. It is simply not physiologically possible for the human body to acclimatise to atmospheric pressure at that height. As a consequence, every minute spent above that height makes the body weaker and increases the chances of injury or death. To stay alive, you have to get up and get out of there as quickly as possible. That's why the line of climbers waiting in a queue is so quietly deadly. Every pause while climbing up increases the risk that you're not going to come back down again.

At its summit, the atmospheric pressure on Everest is a third of what it is at sea level. Reduced oxygen increases the risk of heart attack: once your oxygen dips below a certain level, your heart rate soars to over 140 beats per minute to try and compensate. Then there's the greater risk of having a stroke: the acclimatisation process results in the body producing more haemoglobin (a protein that helps carry oxygen from the lungs). But haemoglobin, which is found in red blood cells, also thickens the blood, making it harder for the heart to pump it round the body, which in turn can lead to a stroke.

The brain, too, struggles with the lack of oxygen. Reduced levels in the brain impairs the ability to think: your mind is slower to process things, and your judgement becomes warped. You can easily forget to clip yourself back onto your safety line, or you can make the fatal decision to press on, when your mind should be telling you turn back.

The lack of oxygen can lead to the brain swelling, which leads to high altitude cerebral edema, or HACE. This results in nausea and vomiting in milder cases and psychosis and delirium in more severe ones: there are stories of climbers starting to talk to imaginary people or trying to take their clothes off. Your body becomes weaker, just when it is being challenged the most; you can't sleep, you lose appetite and your muscles begin to waste.

Everyone on the expedition had oxygen, and our rates were continually monitored. The expedition, in fact, supplied more oxygen to its climbers than any other group on the mountain. But even with those increased supplies, you can still only breathe in as much as the Sherpas can carry up with them. At one point, Fred was told by Lukas that he had to stop using oxygen at the rate he was using it, or he'd run out. The oxygen we had at our disposal was our lifeline, but it's important to understand that it wasn't anything like the levels we had down at sea level. It kept us alive, but it didn't stop the struggle.

Lukas had explained to us in the talk at North Col that if we got injured, we wouldn't be able to get medical assistance to get us back down. But I hadn't really thought about what happened to those left behind. While the Sherpas spend the weeks following the climbing season tidying up the rubbish and detritus on the mountain, the bodies of those who have died are simply too big to be moved. Because of the low temperatures, the mitochondria stop firing, and the liquid in the bodies freezes solid. They essentially become preserved. That was how Mallory was able to be found, all but pristine in the clothes he'd been wearing nearly 100 years before. It was an emotional moment for me when I passed the point where he fell. His body was lying a couple of hundred metres off the main route, from where he slipped. It wasn't possible to see his body from where we were, but I could see the drop and wasn't about to lean over for a closer look.

At times, it can feel as though you're walking through some sort of climbing graveyard. The Chinese have done their best to move as many bodies out of the way as possible, but even so, you can see a hand sticking up here, or a foot poking out of the snow. And there remain many more bodies nearer to the main path, more recent deaths where there hasn't been time to get them out of the way. The bodies are garish and ghoulish – garish in their brightly coloured climbing gear, and ghoulish in how they've frozen as they fell. The arms are contorted, the necks twisted, the kneecaps broken.

The higher we climbed, the more exhausted I felt. That was partly because of the terrain, but also because the thinness of the air made every footstep that bit more difficult. I felt myself relying on the oxygen more and more, gratefully sucking it in at every opportunity. At one point my oxygen tank needed replacing as it was running low. I needed to stop, and sat down as my Sherpa went to work. There was a shape in the snow to my side. As I glanced down, I realised I was perched right next to another body. I could see their face, icy white and frozen. Someone, thankfully, had closed their eyes, but that was about as much dignity as this climber had been given.

I took a deeper suck of oxygen as I set off. A short while later, a cooking cylinder came crashing down from higher up the mountain. It flew past me, less than a metre to the side, hurtling with such speed I could hear the whoosh of the air as it flew past. I braced instinctively: is it going to explode on impact? But with a thud, it hit the rock behind me, and carried on spinning and speeding down the mountain.

As with the dislodged lump of granite in the tent at Camp III, if that had hit me, I'd have been killed – even if not immediately, then injured enough not to be able to make it back down the mountain. It increasingly felt as though I was using up my lives. How many times could I continue to stay lucky? I was used to being in charge and in control. Here, though, whether or not I got back alive felt increasingly in the lap of the gods.

• • •

There was a striking moment on that climb between Camps II and III when I became aware of just how high I had climbed. I looked down and below me; all I could see was a carpet of cloud, the lower slopes disappearing into a soft blanket of white. Then I looked across the sweep of the Himalayas at the other peaks in the mountain range – Cho Oyu, Lhotse, Makalu, Kangchenjunga – that were also recognised 'eight-thousanders'. There were only fourteen mountains on the planet above 8,000 metres tall, and here I was, looking down at them from where I stood.

Then I saw a plane, a jumbo jet, fly across at eye level. That felt disorientating: large planes normally fly at an altitude of around 10,000 metres, and not even I was that high. Maybe it was flying lower than normal. Maybe the bend in the earth created the illusion that I was stood higher than a jumbo jet. Either way, it wasn't an experience I would forget in a hurry.

Beyond, I could see the curvature of the earth. That was a strange, bewildering experience, that I was high enough to see the bend of the earth and the blues of space, all while still having my feet on the ground. The sight was both beautiful and humbling. Standing there I had a deep sense of both the power and majesty of nature, of Mother Earth, and also how small and infinitesimal I was as an individual. Seeing that sight reminded me of why I'd decided to do this climb in the first place. But it also put my attempt in perspective: one man by himself could only go so far in making a difference. Even as I was climbing, the carbon footprint of the plane flying past showed what I was up against.

Looking down and across was nothing compared to looking up. Closing in on Camp III, I was now close enough to follow the route beyond and see it in focus. I could trace most of the route of the north-east ridge, which began a few hundred metres above Camp III. I couldn't quite see the First Step, which was hidden from view, but I could make out Mushroom Rock, a point of the ridge where oxygen is replenished, and then on to the Second Step. They looked scary as hell. Beyond that, I could see what looked like two series of large drops. That was the Third Step: the last of Three Steps before the final haul up through the steep snowfields to the summit. I knew that the Second Step was the most difficult of the Three Steps, but to the naked eye, the Third Step looked all but impossible to climb.

• • •

I was both pleased and disappointed to make it to Camp III. I was pleased because I had managed to make it there in one piece. The scramble up had included traversing a ridge, which was little more than a straight slab of rock to get up and over.

A breathtaking view from Mount Everest.

I still wasn't quite sure how I'd managed that – it felt as though it was against the laws of physics. If that wasn't hard enough, trying to do so in crampons felt crazy. There was nowhere to get a grip and my feet felt as though they were sliding about all over the place. At times, it felt as though there was nothing holding me there apart from the support rope, and I was just hanging there with a 2,000-metre drop below.

I was disappointed because Camp III wasn't really a camp at all. In my mind, I'd been expecting the land to level out a little, the spot for the final camp chosen because it was somewhere vaguely suitable to pitch a tent. The only upside was that we weren't going to be there for long. We arrived about 4 p.m., which was just enough time to get a few hours' rest before the final assault began at nine that evening.

Lukas put Fred, Marco and myself in the same tent, with a telling-off for Fred about how much oxygen he was using up. Maybe it was because of that, but Fred in turn had become cross with myself and Marco. Before setting off, we'd hatched this plan to get into the *Guinness Book of Records* for playing the highest ever game of backgammon on earth. We'd contacted them to find out what we needed for the record to be officially recognised, and they told us that we needed to play the game in front of at least two independent witnesses. The Sherpas were set up to do that.

But when we got to North Col, the Sherpa had taken the backgammon kit out of his bag. It was a really nice set, a roll-up leather one, but even though the kit was travel-light, the Sherpa decided he didn't want to carry any more weight than he'd needed to, and had thrown it out. Without the kit, Marco and I tried to play our game on Marco's iPhone instead. But we couldn't get it to function – the iPhone wasn't working properly at that altitude, and every time we tried to make a move, it wouldn't go through. Although we were blaming the phones, the bigger block was our own mental capacity. Because of the altitude and the effect of the lack of oxygen, I could feel myself struggling to do even the simplest of tasks. After watching us try and fail to get the game up and running, Fred could take no more.

'Guys,' he said in exasperation, 'this is the most important thing we've ever achieved in our lives, and you're trying to play backgammon? What's wrong with you? We've got to start climbing again in a few hours. Get some rest.'

We settled down after that, but none of us got any sleep. It was difficult to go down when you knew that if you stayed overnight here, you'd die. It was hard to get out of your mind that for every minute you were there, your body was eating into itself. It was a strange stop. I couldn't really eat very much; I had some soup, and some water, and that was it. I was also aware that by this point, I wasn't going to the loo. I remember Gabriel trying to go and being unable to do so. I didn't even attempt to. In total, I didn't pee for thirty-seven hours. That couldn't have been good, but I didn't want to stop and think about what it might mean or what was happening to my body.

• • •

Nine p.m. arrived before I knew it. It was pitch black inside and outside the tent. From now on everything was done by torchlight.

In some ways that ramped the challenge up further, but it had its benefits as well: at least now I couldn't see what lay ahead, or underneath if I fell.

It was time for that final double-checking of the kit. I sorted my batteries out for my head torch. I had a new set of socks to put on. I'd kept them in a Ziploc bag to make sure they stayed dry. Wet socks would freeze at these temperatures; that was how you lost a foot. I was tired and could also feel the effects of the altitude on my brain – it must have taken me about ten minutes to put them on.

Marco and Fred were ready ahead of me, so I pushed them out of the tent and told them to get going. Then I panicked about the water. I had one of those thin plastic bottles that marathon runners use. I thought I'd been clever, getting something so light-weight. But I forgot that all the water up here is boiling water – because of the low temperature, any less and it would freeze. I poured the boiling water into my plastic bottle and suddenly thought, this is going to melt through. But the plastic held. Maybe it would be OK, I thought. But then I put the bottle in my summit suit and the heat from the boiling water started to make me sweat. That was a no-go – again, the moisture would freeze, and I'd be in serious trouble. I took the bottle out and the water froze imme-diately. Fuck. I put it in my Sherpa's bag. Maybe that would be useful later. The only water I had left was the small amount in the Thermos I'd been carrying. I wasn't meant to have a Thermos, but thank goodness I did. The 300 millilitres of water I had left there was all the drink I had to see me through.

By now my Sherpa was getting agitated and telling me to hurry up.

'We've got to go,' he said. 'It's getting really crowded. If we don't move, we're going to get stuck.'

Outside the tent, there were torchlights and people everywhere. It did not look promising. This was the worst fear of the expedition – that because of the short window, there would be too many people attempting the summit push at exactly the same time. Not only that, but there were clearly people you didn't want to get stuck behind. One of the first people I saw when I climbed out of the tent was an Irish guy. He was out on his legs, looking all wobbly. This is nuts, I thought, and went over to him.

'Hey, I'm not sure you should be attempting this,' I told him. 'You don't look right.'

'I've got to,' the Irish guy replied. 'It's my seventh summit. I can do this. I have to do this.'

I could understand that instinct and that urge. To turn back when you were so close was a difficult thing to do. What did I know about what lay ahead? He clearly had much more experience than I did. But I tried anyway.

'I really don't think you should,' I said. 'You need to turn back. You're not going to make it.'

'No no no,' the Irish guy continued. 'I know I can do this.'

The Irish guy wasn't the only one in a state. There was a Ukrainian woman who was with three male climbers. They were almost dragging her: she was frantic, shouting in English that she couldn't do it, that she didn't want to go. There was a

Chinese woman with a similar group who seemed to be of the same mindset. The climbers she was with were all but carrying her up.

I checked my headlamp. It was working.

'How are you feeling?' my Sherpa asked.

I looked around at the madness and in a strange way, it calmed me.

'I'm OK,' I said. 'I'm good.'

'Then let's go,' the Sherpa said.

And we started to climb.

For the first part, I had my Sherpa in front of me, clipping me on and off the safety rope, and Rupert, our Austrian guide, behind me. The route out of Camp III was pretty much all granite and I gripped the rock with my crampons as best as I could. Ahead of me, I could see a line of head torches up the ridge, like a row of fairylights. My own torch lit a small circle of ground in front of me, like a small spotlight. That helped me to focus, allowed me to take the climb one step at a time.

My Sherpa wasn't hanging around. Every time we got to someone ahead who was moving slower than us, he wanted to overtake. That was a scary manoeuvre: I had to clip off the safety line and on to the Sherpa. My life was in his hands – and his in mine. I was aware of how much lighter he was than me. If I made a mistake, I wasn't sure if he'd be strong enough to hold my extra weight. If you're overtaking someone while running, down at sea level, you have to move that bit faster to get past them. Now, think about doing that on the side of a mountain, at a 60-degree inclination, when you're struggling to breathe. It wasn't as though once you'd gone past them you could catch your breath. If you did that, then they'd have to stop and wait for you to recover. Which, since you'd just overtaken them, wasn't good form.

The First Step is a couple of hundred metres above Camp III. Now you're just short of the ridge and can start to feel the buffeting of the wind from both sides. Once we were over, we were going to be completely exposed. And here was where we hit the first bottleneck. I saw the Chinese woman who was panicking at Camp III and we managed to overtake her. But when we got to the First Step, the Ukrainian woman was ahead of us, trying to get over. There was no way to overtake at this point. All we could do was stand there and wait.

Now it was the turn of my Sherpa to get worried. The minutes ticked by. Ten minutes, twenty, thirty. This was eating into our time, and into our oxygen. I looked back down the mountain and could see an ever-lengthening row of head torches behind me. People were beginning to shout and tell us to hurry up. I could understand their frustrations, but wasn't sure if it was helping, or just panicking the Ukrainian woman still further. Even so, I could feel myself getting colder and colder as I stood there, waiting.

After forty minutes of waiting, the Ukrainian woman was finally up and over. Our turn. The First Step isn't an easy climb in the best of conditions. At night and at altitude it is challenging. I did my best to follow my Sherpa up. There's one section towards the end where essentially you're having to haul yourself up and across the final gully to get onto the ridge. My Sherpa manoeuvred himself up onto a rock on the side and then built up momentum to hurl himself across and on to the other side. I took one look at what he did and thought, there's no way I'm going to do that.

I tried to get over my own way. I could get my knees above the rock, but it was flat and high, and once I got one leg over, there wasn't enough room to place the other. I was reaching with my arms and pulling on the ropes to get across. I could feel the pressure of the queue of people behind as I continued to struggle. In the end, someone put me out of my misery and gave me a shove up from behind, a hand on my bum giving me that extra bit of momentum and height that I needed.

I was up and over and onto the ridge. Woah. I could feel the blast of wind on both sides now and was grateful for the safety rope. But though we had passed the step, the pace didn't pick up. There were too many people ahead, and not really any places to pass now. Once again, the Ukrainian climber was slowing everyone down.

We paused to change my oxygen tank. This was the small ridge known as Mushroom Rock, and I sat down here while the Sherpa sorted things out. I leant back and put my hand down on what I assumed was rock, but I felt something odd under my fingers. When I looked down I nearly freaked. It was a dead guy's face, sticking out of the snow and staring up at me. With my head torch, I followed the shape of the rest of his body: his arm was twisted and his leg bent in a way that was clearly broken. I felt sick.

Rupert came and sat with us.

'How far have we gone?' I asked.

Rupert, to give him his due, was straight with me. 'We've only just started,' he said. 'We're probably getting on for a quarter of the way up.'

That did it for me. With the effort of getting over the First Step, the coldness from the waiting and the sight of the dead man's face, I started to panic. I remembered the Irish guy at Camp III I'd warned off climbing. Who was out there who was going to say the same to me? I knew we were going slower, much slower, than we had planned. I was beginning to worry I wasn't going to make it. A voice in my head was telling me that the sensible thing was to turn back.

'I'm not sure I can do this,' I said to Rupert. 'I think I should turn back.'

I blinked in the beam of Rupert's torch. 'OK,' he said. 'Talk me through what you're thinking.'

When I finished explaining, there was a pause. 'I hear what you're saying, Hakan. But I'm afraid at this point it's actually going to be safer for you to carry on than to turn back.' Before I could speak, he held a hand up for me to continue. 'Do you know how many people were behind us on the First Step, waiting to get up? To get back down, you're going to have to get round them. If you wait here for them to pass, you're going to freeze. If you try and get past them, you're going have to clip out of the safety rope to do so. I'm sorry,' he said. 'I know this is tough, but the only option now is to continue on.'

I gulped. This wasn't the answer I'd been hoping to hear. 'Come on,' Lukas said, leading the expedition. 'Let's keep moving.'

We continued along the ridge. As we were approaching the Second Step, I could see the queue of lights stretching out in front of us. Another wait. This time, it was longer: it must have been an hour and a half we stood there in the dark before we were able to attempt it. The waiting was tiring too; it wasn't as though you were stood on a flat piece of ground. You were still on the side of the mountain, clinging on, hoping and waiting.

At one point the climber in front of me changed the batteries in his torchlight. I watched as he slipped his gloves off, got the fresh batteries out of his pocket, removed the old batteries and clipped the new ones in place. He had his gloves off for less than thirty seconds, but because he'd had his oxygen turned down low – in order to save it while we were waiting – it meant that there was little oxygen in his system. We heard later that he'd got frostbite, and he went on to lose all his fingers.

My feet were beginning to really hurt; I couldn't tell if it was the cold or if it was blisters. The way frostbite works is that you get severe pain in the affected area. You don't know what the pain is, and then suddenly that part of the body goes numb. That's when you know you've got frostbite. By now the temperatures were crazy – minus fifty – and I became convinced I was losing my toes.

I was getting colder and colder. I turned back to Rupert behind me. 'I don't think I can do this,' I said again. 'I want to turn back now.'

We discussed it again. If it had just been me by myself, I'd have gone. But once again, the guide talked me out of it. 'It's the repeated clipping out of the line to get past everyone coming up,' he said. 'That's how most people die.'

I was stuck. I really wasn't enjoying this. I hated the fact that I had no choice but continue on.

• • •

Maybe it was fortunate I couldn't see the Second Step in the dark. Standing there for ninety minutes, looking at what I was about to attempt, would probably only have freaked me out further. The same would go for the drop. The drop was as large as at any point in the approach – 2,000 metres down if you fell.

Finally, it was my turn. I cannot express how grateful I was at that moment for all that practising in the Istanbul quarry. The first part of the Step is a straight climb up a slab of rock. Then there is a connecting trail of three ladders that the Chinese put up. The third one of these is vertical. Even that only takes you so far: once you get to the top of it, you still have to haul yourself up and over to complete the climb.

Practising climbing in the dark paid dividends. For all my concerns, I could feel my confidence from the training. Once I was on the ladders, I could feel the quality of them. Back in Istanbul, my crampons had cut the ladder to shreds, leaving holes on every step. The Chinese ladder was reinforced and strong as anything – as immaculate as when it had been set up back in 1975. Who says that the Chinese don't make quality? There were other differences too, that were less helpful. My practice ladder was set up from the ground and so had a bit more give and a bit more slope to it; the Chinese ladder was sat flush against the rock.

Then there were the ropes: so many had been laid here over the years that it was easy to get tangled up in them. You had to really concentrate to make sure you were clipping in to the right one. There are times where you have to trust the rope is doing the work and taking your weight. If you clip wrong, then it's going to snap and it's all over.

For all the ropes that were there, there was one that I didn't have. Back in Istanbul, I'd practised with a fixed line and then a second safety one. Here, there was just the single line to work from. Clipping yourself on and off the line is not easy, but especially when you're cold, when your brain is slowing down from the lack of oxygen, and when you're wearing a pair of what are essentially huge thick mittens. At moments like that, when you were unclipped and clinging on, you literally had your life in your hands.

As I climbed, I was aware of the bodies on both sides. They'd flash up in the torchlight as I twisted across and leave me shuddering. What made it worse was that I could see from their clothing that these were people who'd died recently. That sense of the stakes sharpened my focus. Then there were the climbers below and above: people behind wanting to get up, and people ahead, desperate to get down. I was aware I had their lives in my hands as well. There were shouts of encouragement from both queues, but always with an edge in their voice. Less *you can do this*, and more *get on with it*.

The Second Step was tough, really tough. But that toughness was mitigated by my training back in Istanbul. The higher I got on the step, the more my confidence grew. *I can do this*, I thought. *I'm going to do this.* And step by step, rung by rung, my itch to turn back and go home began to fade away. For the first time I thought, *I'm actually going to do this. I'm going to make it up. I'm going to climb Everest.*

As I hauled myself up and over the final ladder, I stood. It was early morning now, and the night was beginning to give way. To begin with the mountain was wrapped in mist and fog, the peculiar blue of first light. But as we left the Second Step behind, continued along the ridge, and ascended the Third Step, that fog and mist began to peel away. Dawn was breaking, the sun was ripping through the fog and all those early morning colours of yellows and oranges on the horizon were sinking in. I was double-blinking in the light and in the brightness. I was now on the final white snowfield to the summit. Ahead of me, I could see a small queue of people waiting on a final pull-up. My Sherpa turned back to me and gave me the thumbs-up.

This was it, the summit of Everest. It was one of those moments where all the pain and misery just peeled away. The rush of release at achieving my goal was almost overwhelming and I had to take a moment to steady myself. As I joined the back of that final small queue, I took a look around. I'd made it through the night and the day on the other side was just spectacular. On all sides, I could see for miles and miles. I took a deep breath, or as deep a breath as you can take when you're on restricted oxygen supply at such altitude.

In every sense I was, and felt, on top of the world.

THE BUSINESS OF
CLIMATE CHANGE

'*Some customers want to know what Interface is doing for the environment. How should we answer?*'

In 1994, Ray Anderson, the CEO of Interface, a US carpet company, found a handwritten note on his desk, forwarded on to him from a sales rep on the west coast. Anderson had founded Interface back in 1973 based on what was then a new concept for the office: the carpet tile. In 1969, he'd come across these on a visit to England and realised how versatile and adaptable they were, compared to traditional wall-to-wall office carpeting. He decided to export the idea back to the US, where it was yet to take hold. Twenty years later, he was the head of a thriving company with 5,000 employees and a strong market share in a competitive industry.

In his memoir, *Confessions of a Radical Industrialist*, Anderson confesses that he hadn't considered the environment much in terms of his business: 'it didn't bother me a bit that Interface consumed enough energy each year to light and heat a city. Or that we and our suppliers transformed more than a billion pounds of petroleum-derived raw materials into carpet tiles ... so what if each day just one of my plants sent six tonnes of carpet trimmings to the local landfill? ... That's what landfills were for.'

But Anderson hadn't enjoyed huge business success without listening to his customers, so when that question appeared on his desk, he knew he needed to find an answer. He decided to set up an environmental task force within the organisation to help shape the response. But then the person heading that task force asked him to launch it by giving a speech about his environmental vision. Anderson had no vision. 'We comply with all the environmental laws' was the sum of all his thinking on the subject.

Anderson came across a book – *The Ecology of Commerce* by Paul Hawken – which appeared on his desk 'as if by pure serendipity'. Hawken's book had a simple but straightforward premise: it was business that was primarily culpable for the devastation of the planet, but it was also business that could play the key role in stopping it. One passage struck Anderson in particular, where Hawken recounted the story of St Matthew Island, off the coast of Alaska. The island was used as a radio station during the Second World War and twenty-nine reindeer were introduced, as a sort of emergency food supply. There were no predators on the island, just plenty of willow bushes and lichen for the reindeer to feed on. A decade later, the game biologist who'd imported the reindeer returned to the island to find the population had grown to 1,350 reindeer. Six years later, the biologist returned again, to find the reindeer population had swelled to 6,000 and that the lichen and willow bushes were beginning to look depleted. Three years later, the biologist visited the island once more. The last of the lichen had disappeared, together with the majority of the reindeer. The population had shrunk from 6,000 to 42.

'I've often described the moment I finished reading that story as a spear in the chest,' Anderson recalled in his memoir. With the parallels to what mankind was doing to the planet clear, Anderson 'felt it was an epiphany, a rude awakening, an eye-opening experience.' Emboldened, Anderson stood up in front of his environmental task force and told them Hawken's story. He told his colleagues that the company was going to change. He wanted every carpet tile the company made to be from recycled or renewable material. He wanted all the waste the company produced to be biodegradable.

The immediate result was stunned silence. The changes that Anderson was proposing were huge for the business. There was no source to buy recycled carpet from. Pretty much all of Interface's manufacturing process involved fossil fuels: nylon via oil, PVC and bitumen, not to mention the coal, oil, gas and diesel used in running the plants and transporting the goods. But Anderson was convinced, and set the company on a mission: his 'mission zero' was for Interface to become carbon neutral by 2020. By the time of his death in 2011, Interface had reduced its fuel consumption by 60 per cent, its energy use by 44 per cent, its greenhouse gases use by 82 per cent, its water use by 73 per cent and the amount going to landfill by 67 per cent. At the time, rather than dragging the company under, revenue had grown by two-thirds. Between 2003 and 2011, Interface made 83 million square yards of carpet using a production process with zero environmental costs. It was a remarkable story of change, started by a simple question from a customer, but driven by a leader with the desire to make a difference.

. . .

In this chapter, I want to turn my attention to the business, financial and economic side of climate change. Making a change to a company like Ray Anderson did is not an easy thing to do. It requires leadership to take both colleagues and shareholders with you. I was reminded again of what Mark Evans of Camston Wrather had told me about the clash between operations people and sustainability folks.

If, like me, you run a business, then you'll no doubt have experienced such a conversation yourself. And it might not be the operation guys in your work that are the stumbling block (my own operation guys, I should state, are fantastic!), but I'm sure you can think of someone in-house who can act as the blocker – those people who still think environmental change will lead to increased costs and less sales.

Amir Kfir is a psychologist, organisational consultant and management change leader who I have worked with for a number of years. When I asked him for his opinions on this subject, he told me about Anita Roddick, who he worked with for several years. In the same way as Ray Anderson rebuilt his company on sustainable values, so Anita Roddick launched and ran her business, The Body Shop, on a simple set of central principles. Launched in the

UK in 1976, The Body Shop's plan was to sell beauty products that were ethically sourced, cruelty-free and made of natural ingredients. From her first shop in Brighton in 1976, Roddick built the business up both nationally and internationally, before selling it to L'Oreal in 2006 for £652.3 million (in 2017, L'Oreal sold it to Natura for £877 million). As Amir described based on his time working with her, Anita Roddick was someone else whose vision drove the business forward.

'Anita was the creator of what is called social responsibility and the whole notion of CSR. Corporate social responsibility really is the creation of Anita more than anybody else. Anita said, "We will not test on animals unnecessarily." She said, "Trade not aid." She said, "There are people, the indigenous of the Amazon, who need our support, we will give them an access into market and we will give them exposure." She put every store of the 1,300 stores that she ended up with as an embassy of goodwill. Every employee had to dedicate four hours a month working for the community. The trucks that travelled the streets of the UK with her products from one place to another had slogans that were statements of a credo, of an ideology. One of them was, "If you think small things don't matter, try sleeping with a mosquito."'

Amir explained how Anita was determined to do things her way: 'She did not want to spend on advertising. She said, "Our actions will speak loudly." For twenty years she did not advertise. The company became public, which she was very unhappy about because now she had to play to the flute of Wall Street. She used to go to the City of London and talk to the analysts. They started arguing with her about all sorts of policies that she instilled, like the ones I mentioned about trade, not aid. Excuse my language now, because I always get excited when I can have an opportunity to speak the way she did. She says, "You fuckers don't understand anything about what I'm doing. Your head is in your butt and in the numbers. You don't understand what it means to build a company." They loved her because they said what you would say when you saw her: "She's on fire." If you have a leader that is on fire, that is passionate about something and is able to instill and ignite this passion with the people around you, you have something unstoppable.'

Finding a way to take your company with you is important. As both Ray Anderson and Anita Roddick showed, when you have both inspiring leadership and strong values, it is a winning combination. Pushing such policies forward is easier, perhaps, when the company is your own. But it is equally possible to achieve if you come in to a company that is already established. A good example here is Paul Polman and his tenure as CEO of Unilever. Paul took over in 2009, having previously worked for Procter & Gamble and Nestlé. As he later said, 'Coming in from the outside wasn't easy.' When Polman took over, Unilever was in decline: turnover had slipped from $55 billion to $38 billion. Action was needed.

Polman didn't just arrest and reverse that slide, he also sought to change Unilever's policy with regards to sustainability. This most obviously came in the form of the company's Sustainable Living Plan. This set out three goals for the company: to improve the health and wellbeing of more than 1 billion people; to enhance the livelihoods of millions of people; and to reduce the company's environmental impact by half by 2030. Polman's plans revolutionised the company. For eight years running, Unilever finished at the top of the GlobeScan/SustainAbility Leaders Survey for corporate sustainability leaders.

. . .

'What is the biggest challenge in terms of expanding Plastic Bank? Capital.'

David Katz's comments about the difficulty of growing his environmental organisation are ones that I found echoed across the board with almost everyone I spoke to in researching this book. Given the success of some of the leaders mentioned above, not to mention the huge potential in companies offering solutions to help deal with carbon emissions, it seems strange that new firms and businesses find it so hard to secure investment for their ideas.

'This is a fundamental question I have,' Tom Chi told me. 'Why can't we allocate capital more efficiently? Obviously, the environment is the most valuable thing we have, as we die quite quickly without it, and each time we even partially degrade it, it tends to make our ongoing expenses "higher", which means most extraction activities create liabilities, not assets. The kind of economy that could last is one in which repairing the damage to ecosystems and fostering their vibrant health over time becomes part of the normal result of human economic efforts. There are no physics-based or economic-based reasons that we cannot build this economy right now, we just got sidetracked with the ease of blunt extraction.'

Tom ascribed the reasons for this to a number of sources: 'Part of it is being focused on short-term gains. Part of it is lack of leadership at some of the higher levels of organisation. But I also think part of it, just to take responsibility on how the passionate people that have worked in the world of impact business and socially conscious business, is that the sector has missed the message. They brought a lot of solutions to market where it's some variation of the pitch: "Hey, if you pay 50 per cent more for this product, it's 30 per cent better for the environment."

'We're not going to get out of our current situation like that. You can't be raising the price and only partially solving the problem. A number of impact businesses sound like, "Hey, these yoga pants have got five recycled water bottles in them." And I'm like OK, but they cost twice as much. So you're intrinsically a niche product. In our firm, the stuff that we go after is the foundations of the industrial economy and we look to radically improve the unit economics. If you have both better unit economics and better environmental economics, then regardless of whether people buy because they care about the planet, or because they want to save money, we all win.'

Mark Herrema's experience trying to get funding for his AirCarbon project at Newlight echoes Tom's thoughts: 'Initially, in our company's lifetime, there really weren't success stories to point to. And so people thought, "In principle, this is good. But can I actually make money there?" And nobody had a Facebook or a Tesla to point to, to say, "Oh, yeah. See. This is a good space to be." That actually remained true for a long time.'

But Mark has begun to detect a shift in recent years. 'Now we're starting to see some success. I think actually Tesla's been an important part of a narrative shift to say, "Hey, products that hit that middle ground of people and planet, people want them." Products like Beyond Meat and Impossible Foods have shown that there's a big market for environmental, social, and governance (ESG) type products. I would say we're starting to see a meaningful turning point in the big investment groups starting to put real capital towards ESG type products that are beyond just solar and wind. I think we'll see an acceleration of that: as more companies are increasingly successful, more will get in. More technologies will emerge. What we saw in software and apps over the past fifteen years, if we could do the same thing in environmental technology, that would be amazing.'

Mansi Konar is an ocean economist, with particular insight and knowledge on the economics of the high seas. She is the co-author of a fascinating report, 'A Sustainable Ocean Economy for 2050', which showed that every $1 invested in sustainable ocean solutions yielded at least $5 in return. Given these rewards, I was interested to know why capital wasn't naturally flowing to these areas.

One of the reasons, Mansi suggested was a lack of knowledge. 'You're looking at big, substantial returns. When I started this work, I was quite surprised, because having worked in this area for so long, I didn't realise that the scale of benefits would be so much larger than the cost. That got me to think, if I'm not aware of it, then sometimes business investors and entrepreneurs will not be aware of these opportunities. Hence, it won't figure in their business strategy. So we will need to educate people and create the knowledge.'

One of the other issues that Mansi flagged is with market distortions. 'So for example, of the around $35 billion in subsidies given to global marine fisheries each year, about $22 billion goes to harmful subsidies that support unprofitable, large-scale industrial fishing operations that exploit the oceans where it's not optimal, to the detriment of smaller fishermen. In a normal production process, you have inputs such as raw materials that go into a factory that produces outputs. The cost of production is a cost of the inputs, and the process to convert these inputs into your final product. You would be paying wages to your labour, the machinery, the capital that you're investing in, the power that you consume in terms of converting these inputs into the output. Now, think about this whole system for a fishing vessel. Here, the inputs are free. You are not paying for generating fish. That's given to you by the ocean. It's completely free. All you're paying for is fuel, and all you're paying for is wages. If your fuel is subsidised then we are just allowing these vessels to just fish more and to overfish.'

Markets are distorted, Mansi argues, because of 'things that are not properly priced or valued. That's where natural capital accounting comes in: you don't really price all the valuable ecosystem services that your environment provides to you, so it's not reflected in the market price. At the same time, there is some bad pollution that goes into the ocean, but it's not really being taxed. That's a failure as well. The result is an information asymmetry that's happening, and it's distorting market decisions. Sometimes, we're not valuing the environmental good, sometimes we are incentivising the environmental bad outcomes, and that needs to be corrected. If not, then it is difficult to make the right investments.'

• • •

Underlying the question of whether the markets will shift their opinions (and funds) about sustainable products, another consistent theme that came up in the interviews was a deeper one, about whether markets and modern economics were missing something when it came to the environment. Do the models economists use actually reflect the values and costs that the environment and natural resources can bring?

Here's Mansi Konar again: 'The way countries are defining progress is not necessarily the right metric,' she argues. 'We use GDP, but GDP doesn't always reflect the value of these natural assets. The natural assets that are depleting aren't always accounted for in the GDP figures. You could have a country growing and showing economic growth, but at the same time, their natural assets are depleting. So there is a question in terms of how we measure what development. For the UK trading with another country, say, how can these trade agreements take account of biodiversity impacts? How can we do business differently? When you're showing you're exporting a certain good in terms of products coming from a forest, how do you account for the fact that you are reducing your natural asset by maybe exploiting the forest in a way that's not sustainable?'

Over the last few years, a number of economists and ecologists have developed what is known as the 'degrowth' movement. One of the long-standing principles of economics is that growth is good. But however efficient economies might become, the greater the growth, the more resources are inevitably being used. Degrowth theorists instead argue the benefits of a planned slow-down in economic activity, but one done carefully – so reducing the working week, say, rather than increasing unemployment. In 2019, New Zealand adopted a 'well-being' approach to its budget: rather than just focusing on GDP, it brought in other quality of life indicators, such as mental health and child poverty.

Hector Pollitt is an economist at Cambridge Econometrics. He specialises in macroeconomic modelling and studying policy impact assessments: among those he models for is the European Commission. For him, there is a fundamental issue with how a lot of economists look at this issue. 'Economics, in my mind, is definitely part of the problem here. I spend a lot of my time debating or arguing with mainstream economists about how much is it all going to cost? And it's always about costs and how to minimise these costs. Because they take such a narrow view of economics, they don't understand the potential benefits and even the uncertainties in all of this.

'In mainstream economics,' Hector elaborated, 'and I include environmental economics in that, which is, basically, neo-classical plus externalities, it's really led us in the wrong direction. I fully agree that carbon pricing has to be a big part of the solution we have going forward. But if all of these people are saying, "It's the only thing we need," this is pretty worrying that they can't understand, even with all the things you see going on around, that the markets will sort everything out on their own and you don't need to go any further than that.'

Hector does see the potential for change in a younger generation of economists: 'The millennials coming through have a much wider interest in the growth of ecological economics, and increasingly what we're seeing are the degrowth people that say, "Economic growth is the problem itself." This debate is generating a lot of interest, but we have no idea yet what would replace our current economic system.'

Tom Chi argued similarly that economic modelling needed a serious overhaul. Right now it is understood to be the economy vs the environment and regulations vs externalities. This is one framing which has left us with the problems we currently have. But 'if you understand the economy to be a subset of the ecosystem, that means that thoughtful investments in the ecosystem can lead to far stronger economic outcomes than an extractive economy. This also gets rid of the idea of externalities because if the economy is the subset, then all of its pollution is internal to our calculations and must be understood directly.'

Between 2012 and 2020 David Claydon ran Macro Advisory Partners, which he describes as 'a boutique advisory firm at the intersection of geopolitics, regulation security and economics.' This description doesn't quite do justice to the stretch and influence that firm had: it hired an eminent group of thinkers and global leaders to help think about the shape of globalisation.

But the longer the business went on, the more questions about climate change began to dominate. 'I was spending most of my time working with investor clients in financial institutions and I could see that they were enormously confused about how would government regulation and markets, how would their stakeholders shape their approach to thinking about climate questions? And so I spent a lot of time thinking and talking about those issues with them.' This led David to eventually leave the firm to start a new group focusing more on climate questions.

One of the factors that David told me is set to make a big difference in the years ahead is that of carbon transparency. Over the past few years, there has been increasing pressure on firms to both collect their carbon data and release this publicly. This reached critical mass at the 2021 G7 summit in Cornwall, when finance ministers made it mandatory for corporates to report climate impacts and investment decisions. 'We are moving into a period of dramatic carbon transparency,' David explained. 'The way in which carbon transparency will make firms accountable, the way in which investors will shift investments based on sustainability is an enormous lever. Given the boom in transparency and a strong shift in underlying investments, I would not want to be a firm whose balance sheet reflected a two, three or three-and-a-half degree reality. I hope if I can be a bit optimistic, carbon transparency, the carbon alliance and those kinds of shifts in big investment flows can have a really, really strong impact and really can shift us.'

The second growing driver for business behaviour that David cited was that of climate activism. 'Stakeholders pressuring corporate management has clearly gone parabolic through the Royal Dutch Shell ruling.' This was a ruling in the Netherlands in May 2021 that Shell must reduce its emissions by 45 per cent by 2030, a legal obligation to align policies with the Paris Climate Agreement. David also mentioned the significance of Engine No.1, an activist hedge-fund that has gained seats on the board of ExxonMobile, in the hope of changing climate policy from within. 'Both these cases add to how consequential the agenda is. They load up the pressure on everybody to realise they need to act, load even more pressure on the private sector.'

David explained that, 'A lot of people, for good reason, have adopted an approach to these issues that are very bottom-up. They're very granular, micro. They start at the firm level, they start at the individual level and they work their way up. Equity analysts treat this sector the same, people managing businesses treat it the same. I can understand why, because that's the way it's been. But this is now I think the most important macro event of the rest of our lives. Fitting the bottom-up realities to a macro schematic that pushes us in the right direction, that's a very, very different environment. People will need to look at this issue quite fundamentally differently over the course of the next couple of years. And the people who've had the vision and frankly the wisdom to act, will be rewarded for it. And you'll start to see the positive consequences thereof.'

· · ·

Whether it is your own company or you've been brought in to take it over, it is possible for a leader to change the dial of an organisation's intentions. And rather than deflate a company's growth, it can help to strengthen their results. I know from my own experiences that this isn't always easy, but when done right, it can inspire those in your team and make others want to join.

I believe that as business leaders, we all have a responsibility to define the environments of our organisations and to pursue policies that are both right for business and right for the planet. With good teamwork, even the most challenging of targets are achievable. In September 2020, I was thrilled when Arçelik was able to announce that we had achieved our target to become a carbon-neutral company. That's a big deal for any company our size, but for one in our industry, it's a particularly tall order. We achieved it through the carbon credits obtained through the introduction of our high-energy efficiency refrigerators; we also collect back the old refrigerators, replacing and recycling them in giant facilities, and using the materials from these recycling facilities in our new products. All of this helps to offset the greenhouse gas emissions generated in our production facilities.

It's a great achievement. But our next aim is to push still further and reduce our Scope 3 emissions. Scopes are the standard measurements of greenhouses gases for companies. Our carbon neutrality has been achieved in terms of our Scope 1 and 2 emissions: Scope 1 measures emissions from owned and controlled sources; Scope 2 for the generation of purchased electricity, heat and steam. Scope 3 is a much wider measurement: it involves the measurements of not just your own emissions, but for all phases of your product's life – we're talking about emissions from all the way up the supply chain, including from companies who provide your raw materials; we're talking, too, about the emissions of your products once in the hands of consumers.

Reducing emissions here is clearly a much harder proposition. Apart from anything else, a good 40 per cent or so of the world's top 200 largest companies don't even report their Scope 3 emissions. For us as a business, this will lead to challenges with suppliers: questions of how we persuade them to change, or whether we need to change our suppliers to hit our targets. It also affects our consumers: how do we help them alter their behaviour to bring emissions down?

But hard isn't impossible. Ray Anderson used to like to quote the physicist Amory Lovins, who said, 'If something exists, it must be possible.' Anderson, too, used to describe his goal in pictorial form: a cartoon drawing of a mountain with a stick figure climbing up it. This he called Mount Sustainability, with the summit being zero footprint. Getting to a summit of a high mountain, as I know too well, isn't easy. But as Lovins says, if something exists, it must be possible.

ON TOP OF THE WORLD

At the summit of Everest, you're on top of the world and a speck of dust at the same time. It's an odd meshing of emotions. There's the euphoria and adrenaline rush that you've conquered the highest peak on earth. But at the same time, there's also the awe-inducing realisation of the sweep, strength, power and beauty of the natural world. Against that, despite all your endeavours, you are nothing.

The summit of Everest is high enough, and the view is wide enough, to see the curvature of the earth. That is a humbling experience, to see with your own eyes the shape of the planet we all live on. The panorama stretches on and on, hundreds of miles of the Himalayas and beyond. It's painted with a simple palette: the whites and greys of the mountain peaks and silhouette outlines of the ranges, and the whites and blues of the sky, the clumps and curls of clouds catching and cushioning the mountain peaks.

The only other colours present were the bright splashes of reds and yellows of the other mountaineers, stretching back in line as they wait for their turn at the top. This was the morning of the photograph that went round the world – the shot of that queue of climbers stretching up towards the summit. I stood on the summit ten minutes after that photo was taken. I wasn't part of that queue, thank goodness: that was taken on the southern approach, the way up from Nepal, and we'd deliberately chosen the northern route, from Tibet, to avoid such a situation. You couldn't see that particular queue from where we stood; that final part of the climb to the top is a mixture of false peaks. You think you've reached the peak and then there is another one to go. That queue was obscured by a dip. But there were enough people on the top of the mountain, and enough people waiting on the northern approach, to get a sense of what it might be like.

When that photo went round the world, it drew criticism and condemnation of how such a precious and isolated place could be turned into a tourist trap – the one place in the world you expect to get away from it all, and you're left to queue in a place called the death zone, where every minute longer than you're meant to be there increases your risk of losing your life, and never leaving the mountainside. Here, the response to the photograph suggested, was the height of folly, of human greed at its worst.

But stand back from that judgement, and there is another way of framing that photograph. Because, yes, there is a whole discussion to be had about how many climbers should be allowed on the mountain, and the rise in the number of passes, and the money made in taking people on those climbs – Everest climbs bring in an estimated $240 million to the Nepalese economy, for example.

But beyond this, the underlying reason behind that queue wasn't to do with the greed of man, but with the might of nature.

The peak of Everest is not just high enough that you can see the curvature of the earth. It's also high enough that it is peaking up into the jet stream. Jet streams are cold, fast-moving winds, that usually flow from west to east. They were discovered by a Japanese meteorologist, Wasaburo Oishi, in the 1920s, and have become instrumental in long-distance flights, reducing flight times. If a wind can propel a jumbo jet, you can imagine what it might do to a person on the side of a mountain. We're talking regular speeds of well over 100 miles per hour, lethal and surprisingly loud – like the sound of a jet engine itself, ironically enough. You might think the peak of a mountain is a quiet place for solitude. Not Everest. And not just because of the other people. Nature's voice is loud and clear as well.

For the majority of the year, Everest sits in the centre of one of these jet streams and is impossible to summit. The highest wind speed recorded on the mountain, in February 2004, was 175 miles per hour. On the universally used Saffir-Simpson wind scale, a hurricane is considered Category 5, the highest category, if the wind speed rises above 157 miles per hour (there is, as yet, no Category 6, above 200 miles per hour, but with climate change, the chance of this occurring is growing ever more likely). Over the winter, the top of Everest is battered by hurricane-force winds three days out of four.

There are only two times of the year when the wind drops – at the start and end of the Asian monsoon, which takes place in

May and September. In both instances this creates a short climbing window that makes summiting possible. But while the winds drop in September, fresh snowfall rules this out as a possible climbing time. Which leaves just the window in May for climbers to make their way up. Even without the extreme wind, the temperatures remain unforgiving: around -25 to -30°C during the day at the summit, and down to -40°C or lower at night.

How long this window lasts for is down to the weather. In 2018, the conditions were as good as they could get. There was the right amount of snowfall which fell at about the right time, freezing at a decent depth to be able to walk on. The window to reach the summit was eleven days: in total, 800 climbers and Sherpas reached the summit.

In 2019, the conditions were about as bad as they could be. Nepal had its highest level of snowfall in forty years. It even snowed in Kathmandu, which was almost unheard of. As for the window to summit the mountain, it shrank from eleven days to just two. Two. So rather than the record number of climbers – 891 – being spread out over a week and a half, you were left with the bulk attempting to do so over two days. That was why there were queues. But the length of the window was nature's decision, not ours. It offered the briefest of let-ups from the jet stream, before it flexed its muscles once again.

• • •

The top of Everest is the size of two ping pong tables. There's a rope around the side of it, which you're clipped onto, to offer you protection from the sheer drop on the sides. How much protection that actually gives you, I don't know. A false sense of security, I suspect. It's not a place for quiet contemplation: with other climbers and Sherpas, there were almost twenty people on top when I summited.

Elbows are everywhere. It's all a bit aggressive: everyone is so pumped for having made it that they're quick to push back, fight their corner for that extra bit of space. I was no different. Anyone who was jabbing me, I was jabbing them back, trying to gain myself a bit more room. Beneath us, on both sides, were queues of climbers, waiting to take their turn on the peak. They're wanting you down as quickly as possible. So are the Sherpas: having got you to the top, all they want to do is get you back. And in the midst of all that, you're trying to soak up the moment.

I summited with Marco and Fred. That had been the original plan, but it ended up happening by accident rather than design. On the way up, they'd been quicker than me. Gabriel, the other main member of our group, had summited about forty minutes before us. Fred and Marco had been ahead, too, but in the bottlenecks on the way up, I'd managed to overtake people, and catch them up. To have them both at the top, to share the moment with them, felt fitting.

I wanted to take some pictures. I had all this kit rigged up to do so – batteries inside my snow suit with cables going out to the cameras. But when I tried to get them to work, they were frozen. I was scrabbling round, looking through all my gear to see what was working. I took my gloves off – a stupid, stupid decision. That cost me the edges of my fingers to frostbite: for about a month

At the summit with Fred and Marco.

after the expedition, those tips were numb and rotten. They stank to high heaven.

The only bit of kit that was working was a Huawei phone I had. Thank God. I pulled out various flags I wanted to hold up – FC Barcelona and various teams that we sponsored, and company brands. *Everest – Tackle Climate Change Now* was the banner that meant the most to me. That was the reason for the trip, and why I was there. I held it high above my head, my gloves down hanging off my sleeves on ties.

Then I took an elbow in the ribs from a Chinese climber. He was shouting and swearing at me in Chinese for getting the flags out. This is China, he was shouting. I thought I was being respectful; I wasn't unfurling a national flag or anything like that. And so I stood my ground, swore back at him in Chinese. *Calm down, for fuck's sake.* It was pressured up there, but I wasn't going to back down. I had one last flag I needed get a picture of, a plain green one that I knew I could use to photoshop in other images. The elbow in the ribs from the Chinese climber wasn't the last one I took for that. Later on, Marco and Fred would be pissed off with me for that picture as well. I was the one person with a photo of a sign with a message to my wife. They'd brought nothing up to show to their loved ones. *Where's my picture?* They were ribbed back home. *Why couldn't you be more thoughtful, like Hakan?* They made sure to let my wife know that her message had been photoshopped in. And I got an elbow in the ribs again for my troubles.

I put the flags away, finally saw sense and put my gloves back on. I took a final moment to try and take in the view. It was

a strange feeling, almost eerie. The space and scale is so different, it was difficult to get your head around. We're so used to living in cities and enclosed spaces that to experience a sweep like that messes with your perspective. When you see the world's curvature, the thin line to outer space and the deep blue coming down and fading out from that, it changes how you think, how you see. I couldn't take it all in, couldn't really make proper sense of it all until later.

All the while, the euphoria of summiting was wearing off, fast. It was being replaced with that lurching sensation that I was going to have to make my way back down. All I'd thought about was making way to the top. Now, it was all about getting home. As the euphoria washed off, I could feel a wave of tiredness wash over me. I was exhausted from the effort of having got there. But what I faced now was even harder. It suddenly felt horribly real.

I didn't know it at the time, but there were people in those lines, queuing to get up, who were dying. They couldn't move. Such was the bottleneck that they were stuck. Everyone else had to climb over them to get past. Maybe there was something in the air because of that. But our Sherpas were worried. While I was getting my head around the view, they were watching the weather. And the wind in particular.

Dotted down the mountain were our oxygen supplies. Rather than carry all those canisters up the mountain, they were stored and stashed in set places along the route. Each different expedition had their supplies clearly marked, so everyone knew which canisters belonged to who. But with the wind beginning to whip up, these supplies were being blown off the mountain. Those that weren't blown off had broken free from their moorings, lying around unmarked for anyone to pick up. From further down the mountain, there were messages on the radio that Camp III had also been hit by the wind, with more equipment being blown away. The Sherpas knew they need to move fast in order to secure what supplies were left on the mountain, and to make sure we didn't lose any further equipment. *It's not good*, they said. *We need to get out of here.* And with that, without further explanation, they were off.

That hit home. I was their meal ticket. They only got paid their bonus if we made it back to Base Camp alive. But rather than looking after us and earning their corn, the situation was serious enough for the Sherpas to prioritise getting back themselves, making their way off the mountain as quickly as possible. The Sherpa who'd been with me all the way up, clipping me on and off the ropes, was gone, disappearing down into the distance.

In that moment, among all those people on the peak, I felt horribly alone. And scared.

This was no longer an endurance challenge, an expedition with friends. This was now a matter of life and death.

Of survival.

* * *

It was the flash of wind that got me first. The jet stream might have calmed enough to make summiting possible, but it didn't mean that the wind had dropped. There was a flare of wind, a flick of Mother Nature's fingers, a 100-mile-per-hour gust. That flattened me at a single stroke. One second I was stood up, the next I was horizontal, face down in the snow. If I hadn't had my harness attached to the rope, that would have been me gone. I staggered back to my feet and then it happened again. *Whoomph.* The sheer strength of that blast was both amazing and terrifying at the same time. But mainly terrifying. That sense again of how small I was and how powerful nature could be hit home as hard as I'd hit the deck. I'd achieved what only a few thousand people had ever managed in climbing Everest. But nature seemed to be making its point. How pitiful my achievements were against that might.

I wanted to get the hell out of there. All the excitement and enjoyment and euphoria had been well and truly blown away with that gust. I wanted to get down, but doing so was easier said than done. You couldn't rush. Every footstep in the snow had to be chosen carefully. It was like walking through a minefield. Go too fast and you'd quickly be going faster still, out of control and out of sight before anyone could do anything about it.

I looked around for our team. There'd been a dozen of us together on the top of the mountain. Now we were strung out. The majority of the Sherpas were much further down, disappearing behind ridges and shrinking into dots. Lukas was with them, watching our progress from afar. The only Sherpa behind was the head Sherpa, sweeping up at the back of the group. He had what was left of the oxygen, but with the other climbers panicking and pushing past, he was a way away as well.

As I struggled down towards the Third Step, I could feel my legs buckling, nervous in anticipation of another gust of wind. Behind me were more climbers, desperate to get down. I could smell them – no one had washed or showered in however many days and in the thinness of the air everyone's ripeness rippled down. BO and sweat and shit. And I could hear them, too. The wheezing of their breathing. The shouts to hurry up. To move over so they could get past. Which, when you're losing your nerve, worried about losing your footing, is the last thing you need to hear. It just ratchets the pressure up still further. I could hear the desperation in the voices. *If you don't hurry up, you're going to kill us all.*

I made it to the Third Step. Out of the Three Steps on the north face, this is considered the easiest – just a ten-metre drop down a sheer cliff face from the summit snowfield to navigate. Just. Going up had been difficult enough. Going down was a different level. You had to lower yourself down over the edge, get a grip with your feet, then push yourself off and down. The ropes that you're clipped to are crucial. They're literally your lifeline. But there is no tension to those, nothing to arrest you if you fall. Those ropes and whatever strength you have in your arms are all you have.

There was a bottleneck when I got there, which didn't help with the nerves. As I waited for my turn, I realised I didn't have the right carabiners, those all-important metal clips that you use to attach yourself to the rope. Those were meant to get changed at the summit. On the way up, you use what is called a jumar, which is a clip that works with friction: as soon as weight is applied to it, it tightens. For going up, that's great: it means that if you slip, it stops your fall almost immediately. Going down, however, that's not so useful. At the top, the Sherpas are meant to replace that jumar with a second fresh carabiner; as you switch from rope to rope, you've got the safety of knowing you're always clipped on to one of them.

Because of the confusion at the top, as the Sherpas left in haste, that change of equipment hadn't happened. I was stuck descending without that extra safety precaution. I was going to have to use what I had.

My turn. I went over, as I'd been instructed. But my boots, which were meant to be giving me hold against the rock face, weren't gripping in. In a panic, I realised that I was starting to slide, and couldn't stop. Fuck. I'll turn around, I thought. If I can't use my feet, I'll rely on my arm strength. I'll use that to lower me down. But as I tried to do this, I heard a ripping sound. The rocks were slicing open my summit suit, letting loose the feathers inside that were keeping me warm.

It got worse. As the rip released the feathers, they started clogging up my regulator. I was breathing out rapidly enough with the exertion and pressure of the situation. But with the regulator blocked, I couldn't breathe. As I was hanging there, trying to hold it together, the humidity coming out of the regulator was also steaming up my goggles – if only I hadn't given up my sunglasses to Fred.

I couldn't breathe. I couldn't see. And then, I felt my big leather mittens beginning to slip. My grip on the rope was beginning to weaken. I could feel myself starting to slide. My heart was really beginning to hammer. *Get out of the fucking way!* Someone shouted from above. *You're going to kill us all if you don't get a move on!* Someone else yelled. *We're all going to die up here. Fuck's sake. Move!*

Mentally, physically, I was right at the limit. Then, just as I was about to panic and properly lose it, I winced and cramped as a pain in my stomach erupted. What the … Suddenly, my body felt weird. And then, incredibly, it felt really good. I felt pumped, primed, my veins pulsing. My senses, wow, they clicked into focus. Stronger and sharper and crystal clear. It was like I could hear a conversation a mile off. It was like going from normal TV to ultra HD and then some.

Later, I learned via a conversation with my friend, the esteemed Stanford neuroscientist Andrew Huberman, that what I'd tapped into is something called the sympathetic chain ganglia. This is a bundle of neurons that runs down your spine from the level of your clavicles down to just above your pelvis. In extreme situations, they fire up. It's a bit like how the adrenal glands release adrenaline, but on a much, much larger scale. A sort of military level, rapid reaction force for the body.

The firing of these neurons allows you to tap into the full strength of your body in a way you can't normally access. In terms of your musculature, you're able to access high-threshold motor units that give you extra strength. In terms of your sight and sense of time perception, the neurochemicals give you enhanced focus. It feels as though time is slowing down, because your perception is heightened.

The science behind this phenomenon, sometimes called hysterical strength, is still relatively new, but it's the explanation behind those strange situations where, for example, a woman has the power to lift a car to rescue a loved one, or a person can go toe-to-toe with a polar bear to save their friends. It sounds far-fetched, almost superhuman, but it's actually just the body working at full capacity. Normal people doing sport might use 60 per cent of their capabilities, elite athletes 80 per cent. It is only in extreme life and death situations that you experience what the body is like at full stretch.

That was what I was going through. I could feel the flooding of my body with adrenaline, the adrenal glands pumping it into the blood. That helped with my heart rate and breathing, the extra-oxygenated blood in my muscles and the activation of these high threshold fibres giving me a strength I didn't realise I had. My pain threshold disappeared. The pain hadn't gone, but I no longer noticed it, allowing me to push through in a way I couldn't before. All of this happened in a split second, but the sensation was extraordinary. I'd gone up Everest to witness the power of nature. But I hadn't expected to experience it from within as well.

Time for action. I ripped my goggles off, threw them down. It was bright without them, blinding bright, but at least I could see again. I took off the regulator, sucking in the ice-cold mountain air. I could breathe. I knew I didn't have long – more than a minute without oxygen at that altitude and my brain would be overpowered by pins and needles, and I'd lose consciousness.

The seconds were ticking.

I shut my eyes, calmed down, centred myself.

You can do this.

Turn back around, put pressure on your feet, get a grip.

I gave out a guttural roar. With my new-found strength, it was like my hands had transformed into suction pads. The slipping and sliding stopped. I was back in control. With the crampons on my feet and the rope in my hands, I managed to manoeuvre myself down.

I'd survived.

Just.

As I made it off the Third Step, I put the regulator back on, breathed the oxygen in as best I could. As I did so, I felt over-whelmed. A wash of emotion came over me and I welled up, bursting into tears as the enormity of what I'd done hit home. That pain in my stomach was back – that strange sort of warmth, like I'd shit and pissed myself at the same time. Maybe in the moment I had. All I knew was that I was shaking. I remember looking down at my hands and holding them up as they wobbled in front of me.

Thank God for help. From lower down the mountain, our Austrian guide came back up. He picked up my goggles from where I'd thrown them off, wiped them and put them on himself. *Here, take these*, he said, handing me his sunglasses. I didn't know what to say, but knew what it meant. With that one gesture of generosity, he had saved my eyesight from snow blindness.

The Sherpa who'd stayed at the back of the group made it down behind me. He'd been watching what had happened from above and fixed my regulator as best he could.

'Hakan, you have to get down quick now,' he warned, waving at the sky. Through the guide's sunglasses, rather than my steamed-up goggles, I could see the weather was starting to close in. 'It doesn't matter how. Don't stop, keep moving, go fast. If you don't, you will get left here, do you understand?'

I nodded. After Step Three, I knew exactly what was at stake. I didn't plan on hanging around.

• • •

As I approached the Second Step, I saw the outline of two bodies in the snow. These were climbers who had died on the mountain, left there for years, a chilling, frozen reminder of the danger we all faced. But on the Second Step there was a much more recent body. The Second Step is the hardest of the three: it's about 130 feet in height, with the top section a sheer drop of about fifteen. There is a series of aluminium ladders for you to climb up and down, but they don't reach the top; you have to lower yourself over the edge and feel around with your feet for the top rung.

As I took my turn, tentatively and thankfully finding something underfoot, I was grateful for my good fortune. But not everyone was as lucky as me that morning. Barely ten minutes after I descended the Second Step, a climber called Ernst Landgraf attempted the same manoeuvre. As he'd tried to put his foot on the top rung on the ladder, his footing had slipped. Because he was attached to the fixed line, he'd banged hard against the ladder, then hung still. His expedition leader realised he'd had a huge, immediate heart attack. The climbers around him had rushed to help, but quickly realised he was dead. Such was the crush of

The perils of the Second Step. An unfortunate climber collapsed and died minutes after I descended.

people needing to descend, they didn't have time to take him down or cut him loose. They'd simply pushed him to the side and left him there, hanging. Later, when I saw the photos of his body, hanging limply while other climbers went past, I was horrified. I remembered what had happened to me on the step above and thought, there but for the grace of God.

I pushed on. By now I was making every promise I could think of to get me out of there. Everything I enjoyed doing, whatever came to mind, I swore I'd never do again as long as I got back safe. One item of help and support I turned to at this point were my crystals. What I would do with them, which is part of a practice that I've learned, is that as I was walking I would repeat the things that I wanted to let go of in my life. That I wanted cleansed from my life. During the expedition, whenever I found a suitable spot, I would give a crystal to the mountain. I would say a little prayer, set an intention, and ask the mountain to allow me to come back safely to my family. As well as using the crystals to leave the negative energy behind, I kept others that I loaded with positive thoughts, to help keep me going. I know that some people might be sceptical reading this, but in those dark moments, having those crystals there really helped.

The weather drew in further. I tried to take the descent as quick as I could, but more than once I felt the looseness of the rock giving way underneath, and I had to slow myself before I lost my footing and gave way. The adrenaline rush of getting down the Third Step had completely worn off. Exhaustion echoed in my bones. I was completely spent, could feel every ache and pain in my body.

Marco was ahead of me. From the way he was moving, I could see that he was struggling as well.

'You OK?' I asked, when I caught up with him.

'Oxygen,' he said, the panic rising in his voice. 'I'm running out.'

I'd been careful with my supply, but Marco had been burning through his. I looked around in the hope of seeing our Sherpa with the spare canisters. But he was nowhere in sight. Marco was a friend. There was no way that I couldn't give him some of mine. I had no choice, whatever the risk to me.

Further down the slope I was stopped again. It was a climber clad in black, desperate for oxygen. I felt awful, but I knew I'd already taken a risk in helping Marco. I couldn't give any more out. I shook my head, said I was sorry, and prayed that someone else was able to help him.

'Please!' I heard another shout. This time it was a woman, sat by the side of the main route. Straightaway I could see she was in a state. Her gloves were off, which was not a good sign. She was bordering on delirious, leaning and grabbing towards me, begging me for help.

My water supply was even more precarious than my oxygen. The water I'd had on the way up had frozen and I'd foolishly given it to my Sherpa to carry. All I had was a Thermos flask, which I wasn't meant to have taken due to its weight, but it had slipped into my bag. That had about 300 millilitres in it: that was all I had with me to summit and get me back down to camp.

'I need water. Please. I'm begging you. I'll do anything. Anything.'

Just as I could not say no to Marco, I couldn't leave this woman like this.

'I've got a thermos,' I said, getting it out of my pack. 'But I think it's frozen.'

'No no no,' the woman said, with a renewed urgency. 'There's water in it.' She tried to open it, but her hands were frozen. I took my gloves off and opened it for her. I thought she'd take a grateful sip, but I watched as she gulped down the whole lot in one go. It was like someone at Oktoberfest draining their beer, so fast it doesn't touch the sides. Fuck, I thought. That's all I had. I'd thought she'd just take a mouthful, but now it was gone. Suddenly, my mouth felt dry with thirst.

I helped the woman up. I supported her, attempted to get her moving again, but after a few steps I realised it was pointless. She was in no state to go anywhere. It was heartbreaking. I helped her back down, but had to leave her. Later, I learned that she'd been left by her Sherpa, who'd run out of oxygen. Her guide had eventually gone back up to get her that evening, somehow finding her curled up in the dark. He managed to get her back to safety, saved her life. But there was a price: with her gloves off, she'd ended up losing her hands to frostbite. I knew that I'd been in no fit state myself to help her down, but at least my water had helped keep her alive until somebody else could.

North Col was coming closer. I knew that if I could make it to there, I'd be OK. But all the time, I could feel the pressure of the time ticking on. Not just with the weather, but also with the night drawing in. It had been 7.20 in the morning when I'd summited.

By now it was late afternoon. To be honest, I didn't really know what time it was. I just knew that it was beginning to get dark, and I did not want to be out on that mountain when the lights went and the temperature started to drop.

Just as I was starting to think about the possibility of reaching camp, I slipped. I was trying to clip my carabiner off one rope and onto the other, but as I did so, I got it caught in my summit suit. Before I knew it I was sliding, unattached. In desperation, I grabbed out and caught the rope in my hand. That was all that was stopping me from sliding down and off the mountain. The slopes down each side were so steep, sheer like a ski jump, there was no way I could have stopped myself.

I had that sensation in my gut again and grabbed tighter on to the rope.

'Marco!' I shouted. 'Marco, I need help!'

Marco was ahead of me, further down. In the dying light, I wasn't sure if he had heard me.

'Marco!' I shouted again.

I saw him stop, turn round.

'Help me!' I shouted again.

'What?' he shouted back. 'Hakan? Are you OK?'

'I'm slipping,' I shouted. 'I'm going to fall.'

Marco's English is not always the best, and there was a pause while he tried to understand what I was saying.

'Please,' I shouted.

If he did not understand what I was saying, at least he understood the panic in my voice. He came running up, just in time.

I could feel my strength going, and wasn't sure how much longer I could have held on for.

'It's OK,' Marco said, clipping me in and helping back up.

I breathed out in relief. I'd given him my oxygen, now he'd saved my life in return. What goes around, as the saying goes.

• • •

By the time we made it to North Col, it was 8.30 in the evening. Thirteen hours of pure hell. It was pitch black on the mountainside. I was so exhausted that I'd stopped walking and had regressed to crawling, heading down on all fours, which was about as much as I could manage. In the camp, Fred, Marco and Gabriel were all sat together around a small fire. A chair was found for me, close to the flames. I collapsed into it. None of us had the strength to speak.

The Sherpas wanted us to keep going. That was the order from Lukas. But with what little energy I had left, I shook my head, and so did the others. I was fit for nothing. There was no way I could leave the camp here and walk on through the night. If I was going to die, I'd die here, I thought. The Sherpa tried to convince us again, but then gave up. It was clear we couldn't go anywhere that night.

VOTING FOR CHANGE

In the late 1960s, James Lovelock noticed a strange phenomenon at his home in the English countryside. Lovelock lived in Bowerchalke, Wiltshire, a quiet, rural village in the county's Chalke Valley. Over the years, the village has also been home to William Golding, the Nobel Prize-winning novelist, and the First World War poet Siegfried Sassoon. But for all their literary influence and significance, Lovelock's inquisitiveness would ultimately save the world.

What Lovelock noticed was a haziness in the Wiltshire air. This struck him as strange. Haziness was a phenomenon associated with large cities, like the fumes that enwrapped Los Angeles. The nearest known incidence to Lovelock was that of London, ninety miles away, which a decade earlier had suffered from deadly smogs. Those, however, had been halted thanks to the British government passing the Clean Air Act. For such air pollution to occur many years on, and in the middle of the countryside, didn't seem right. It also seemed something new: Lovelock racked his brains back to his childhood, and couldn't recall ever experiencing anything similar.

At this stage in his career, Lovelock described himself as an 'independent scientist'. Having worked for institutions for years, he now experimented and invented from his home laboratory. One of his inventions was the electron capture detector (ECD), a machine used to detect chemical pollutants. Lovelock's device had been adapted from a failed attempt to analyse cell membrane damage and it was one of science's happy repurposings: the ECD was far more sensitive in detecting chemical pollutants than any other instrument.

Using his ECD, Lovelock began to try and measure the haze, and deduce where it came from. What he discovered was that the haziness was linked to the amount chlorofluorocarbons, or CFCs, in the atmosphere. CFCs were commonly found in such everyday items as aerosol cans and refrigerators. Lovelock then travelled to western Ireland and discovered similar results: two rural, isolated places with ozone levels above recognised safety limits. Lovelock published his findings to little response, but pushed on with his research anyway. He managed to get a place on *Shackleton*, a research ship heading from the UK down to Antarctica. Lovelock travelled as far as Montevideo, measuring the air quality across the Atlantic. Part of the journey followed the route of Charles Darwin's *Beagle*, some 120 years earlier. Lovelock's results were similarly revelatory: having found evidence in the southern hemisphere to match his earlier experiments, he was now convinced that CFCs were in the atmosphere across the globe. As he disembarked in Uruguay, the ship's captain said to him, 'This will be remembered as one of the key research voyages.'

From Lovelock's early discoveries, the investigations continued. Scientists discovered that CFCs reacted with and then depleted ozone in the stratosphere. The ozone served as a protection against the UV light of the sun: the loss of it would lead to effects such as skin cancer, DNA damage and the reduction of phytoplankton. By 1985, scientists had detected a 'hole' in the ozone layer above Antarctica (actually a diluted thinning). Without a sustained response, this would continue to grow in the years ahead.

Despite the growing evidence of the irreparable damage that CFCs were causing, the chemical industry continued to hold firm. Manufacturers clubbed together into the Alliance for Responsible CFC Policy, arguing that the science was unfounded and there was no need for any form of intervention and regulation. To begin with the US government backed them up: the Secretary of the Interior told one meeting that Americans should just wear hats and sunscreen. But three months after the revelations about the ozone hole over Antarctica, the US president, Ronald Reagan, needed surgery to remove a skin cancerous growth on his nose. Suddenly, the linkage for him seemed to make sense, and his administration changed tack.

In 1987, the Montreal Protocol was signed, a global treaty with countries agreeing to phase out CFCs and other substances affecting the ozone layer. Forty-six countries signed the original agreement: by 2015 it had been ratified by the entire United Nations, the only treaty ever to be signed by every country on

Imja Glacier 1956 vs 2007, Himalayas. (Source: NASA)

earth. Today, 98 per cent of ozone-depleting substances (ODS) have been phased out compared to 1990 levels. It is estimated that the treaty has saved 2 million people a year from skin cancer. The damage to the ozone layer has been halted and reversed, with the hole expected to be closed up by the middle of the century.

The Montreal Protocol is now regarded as one of the most successful environmental treaties of all time. From James Lovelock's first investigations to an international agreement two decades later, the threat to the ozone layer, so potentially devastating, has been dealt with. It's a good example of what can be achieved when the world does come together, and how true international co-operation can play its part in tackling environmental issues.

* * *

In this final environmental chapter, I want to focus in on the politics of climate change. Climate change is a global issue that crosses borders to the point that state boundaries can feel almost meaningless. Take the glaciers in the Hindu Kush and Himalayas, for example. These don't just supply water for the 250 million people in the region, but the rivers that flow from there wend their way down into India, Pakistan, China and beyond. The catchment area for these glaciers is 1.65 billion people. That's almost 2 billion people whose livelihoods depend on these glaciers. And they're melting. Even if we manage to limit temperature increases to 1.5°C, a third of these glaciers will still be lost by the end of the century.

The effect of this melting will be increased by river flows and flooding downstream. But later on, the opposite problem will occur as the rivers go into decline. For agriculture, even if the catastrophic flooding expected to engulf whole communities can be avoided, the loss of predictable water supplies will make farming incredibly difficult. The political ramifications of events like these are conflict and migration. At present, there are approximately 64 million forced migrants around the world, leaving their homes because of war, hunger and persecution. The UN estimates that with effects of climate change, the number of environmental migrants could reach 1 billion by 2050.

The melting of the Himalayan glaciers is just one example of many I could give. The rise in sea levels due to melting polar ice caps will put pressure on low-lying coastal communities worldwide. Long rivers such as the Nile, which cross several borders, will also be a potential flashpoint for conflict. Elsewhere, increases in temperature will make some parts of the world uninhabitable. Even if, as a country, you work hard to combat climate change, the behaviour of others may still have a deleterious effect.

As COP26 showed us, just because the science is incontrovertible doesn't mean that action will be taken. David Claydon put the reason to me like this: 'Ask yourself the question, if the science is unarguable and there's an engineering solution, what are the impediments? The impediments are often political. And once they're political, how do we translate engineering and scientific knowledge into policy action? That's really the key question.'

Rather than the world coming together on the issues, David worried that, in fact, countries were splintering into different groups: 'I think the world is beginning to break into three blocks on this. One is broadly a G7 plus, The Democratic 10 (The G7, plus Australia, South Korea and India), which makes up about 25 to 30 per cent of the global economy, and is fully engaged with this agenda. The second block is China. China is doing a mix of fossil fuels and renewables and has obviously changed all its reporting to obfuscate the fact that they continue to build and burn. If you look at it in detail, they're really not doing very much to reduce the total amount of coal being burned. The primacy of growth, in the post-COVID era, still beats reducing coal as a priority.

'And then the third block is the global south. The global south is just doing fossil fuel. There's some renewables here and there, solar in Morocco, for example, and some distributed micro solar in some countries in Africa, but the $100 billion that the COP nations at Paris committed to help emerging countries, the so-called differentiated responsibility of developed economies, that hundred billion, I think has been MIA.'

David also worried about the political challenges of bringing everyone on board with climate policies. 'In developed countries, particularly in Europe, you're starting to see climate policy criticism become a tool of a populist right. Back in the day, migration was their driving issue, and the climate policy narrative was just a scare story. Five years ago, the relative intensity of the migration and asylum issue was ten out of ten, climate policies were at two. Now asylum and migration has probably softened a bit, and the climate policy issue has risen to a five or six out of ten.'

David was concerned at how climate change policies were being pitched as 'win-win': 'we make the changes and we make the investments, we get growth jobs, high-quality jobs and an inclined outcome. I'm concerned that is a bit simplistic and there will be winners and losers. As we get closer to having to make all these changes, we're going to have to have really strong politics to make sure we stick to our guns when the distributional costs begin to become apparent. If we don't frame the politics right, we're going to be in a right pickle.'

Speaking to David in summer 2021, in advance of COP26, he expressed concern that we were heading into the negotiations 'with less co-operation than we might have hoped, and with more risks to the underlying momentum than is necessary.' The arrival of Joe Biden as US president had shifted the dial in comparison to Donald Trump's attitude to the environment, but the 'confetti of advances' came with a risk attached: 'I think the US is probably going to under-deliver on its commitments, and we'll be back to a US that commits a lot but doesn't deliver very much.' All of this, David noted, is taking place 'just as the underlying scientific outlook from the IPCC is deteriorating.'

Another fascinating thinker in this area is David Victor, a political scientist, engineer and oceanographer who has written extensively on the challenges of international co-operation. David is one of the authors of 'Accelerating the Low Carbon Transition: The Case for Stronger, More Targeted and Co-ordinated International Action.' It's a fascinating and practical report.

David is well versed in the politics of climate agreements and in the reasons why they often fail. 'Most of the global climate efforts to date have failed because they were disconnected from facts on the ground,' he had written previously, 'from what governments and firms are willing and able to do.'

I asked David about the Montreal Protocol, why he thought that agreement was successful and why we haven't learned the lesson from that for future discussions. 'Everyone thought Montreal was a huge success, correctly,' David said, 'and then they apply the lessons of Montreal to the framework convention on climate change incorrectly. And thus, we've spent almost thirty years, until Paris, with the wrong framework.'

So why was Montreal a success? 'It was successful because it took the big problem and it broke it down into small units. And then it actively experimented around technologies for each of the small units and clusters. There's a whole machinery inside Montreal that runs almost invisibly. It's machinery familiar to engineers: there are companies out testing things, and some of it works, some of it doesn't. They come back and they do a centralised peer review, and then they go that route and adjust the goals that way.'

David describes this as 'experimentalist learning.' When the original Montreal agreement was signed, the science wasn't completely clear on either the causes or what could replace the ODS. But setting a target to cut emissions sent a signal to firms; they in turn responded and found solutions. With the solutions clearer, governments felt more confident in making the cuts larger. At the same time, the cuts were broken down into individual sectors and individual countries: that 'finer-grained focus' helped enormously. Thirdly, as well as this approach of experimentalist learning, there was also what David calls an element of experimentalist governance. Rather than trying for a global agreement from the off, the Montreal Protocol was signed by interested parties in the first instance and then, as the science became clearer, more and more countries signed until, in the end, everyone did. David argues that just because environmental issues are global problems, it doesn't mean that solutions have to be global from the start. This way round, more ambitious targets might be agreed than for an initial global agreement, with others then compelled to come on board.

The model of Montreal is a useful one, but climate change offers different challenges. The issue of CFCs and the ozone layer was largely focused on one particular problem and one particular industry, which made tackling it easier. With climate change, the issue affects industries across the board, which makes finding solutions more difficult. Nevertheless, David argues that breaking down the challenges into sectors and industries is the way forward, and his low-carbon report separates the issue out into ten main sectors: power, agriculture, cars, trucks, shipping, aviation, buildings, steel, cement and plastics. 'The Paris Agreement is great as an independent general convening authority,' David argues, 'but it's too big to get things done. It's smaller groups of countries and firms where this can happen. Take shipping: it's Maersk, the Danes, a few others, the big shipping dyads, they're the ones who are testing ammonia, testing hydrogen and working on that.'

As well as creating this co-ordination within sectors, David argues for the importance of institutions in overseeing all of this. Apart from plans for decarbonisation, he sees this as the single most important activity for governments to undertake: to strengthen existing institutions, and forge new ones when needed. 'What institutions do is they help those governments co-operate, because they are off doing their own thing. And if everybody goes off and does their own thing, then you just have chaos. The role of institutions is to stabilise expectations and to allow those smaller groups to co-operate.'

• • •

The importance of institutions and international co-operation is one that came up time and again in my conversations. When I spoke to ocean economist Mansi Konar, she made clear how this was a particular issue on the high seas. 'The problems of the ocean, they're not restricted to borders or particular EEZs (exclusive economic zones). You have countries who are managing their own waters, their own sovereign territories, but fish don't really know what these borders are. They're free-roaming. The other issue is that when leaders are managing their oceans, they're very much looking at silos, so whether it's within a particular ocean-based sector or within a particular region. Studies have shown that disjointed and ad hoc management regimes don't really work and have resulted in poor outcomes, such as increased marine pollution or overfishing.'

But Mansi also thought that there were signs of change here. She talked about her work as lead economist for the High Level Panel for a Sustainable Ocean Economy. This group consists of fourteen Heads of State that are demonstrating a new type of global co-operation in building towards a sustainable ocean economy. The countries involved are from across the world, including Norway, Canada, Mexico, Chile, South Africa, Japan and Indonesia. Between them they account for about 40 per cent of the world's coastlines and 30 per cent of the world's EEZs.

In December 2020, the Ocean Panel put forward a bold new ocean action agenda underpinned by sustainably managing 100 per cent of the oceans under their national jurisdictions by 2025. Additionally, they vowed to set aside 30 per cent of the seas as Marine Protected Areas by 2030, in keeping with the 30by30 Global Ocean Alliance initiative. 'Before this panel came together,' Mansi explained, 'there wasn't any panel that had a strong political focus and mandate on the ocean, with Heads of State saying, "We want to do something about this. We want to not only bring recommendations, but drive and accelerate action."'

This echoes David Victor's comments about Montreal and how interested parties signed up first, with others following suit as the evidence became more incontrovertible: 'I think that was the intention,' Mansi agreed, 'that you start with these fourteen leaders, with your champions within your region, in order to bring about a bigger coalition to join in this drive towards action. That's what the ocean panel has been doing. It keeps saying, "The ocean gives us so much – it feeds us, entertains us, connects us, inspires us and powers our success. So let's give the ocean back 100 per cent and join us in this initiative and endeavour."'

For Mansi, such is the size of the issues involved that the answers lie beyond just governments. 'I think we need more visionaries, not only from the sphere of politics. We need it from businesses, we need it from entrepreneurs, coming forward to say how they're going to address some of the big challenges the world faces today. Government can't solve the problem alone? But the prize for solving it is potentially game-changing: 'When you look at the issue of climate change, or you look at the issue of pollution, or you look at the issue of food security, the ocean hasn't really featured in one of these big problems that the decision-makers are thinking about, yet it holds the solutions.'

• • •

One person who knows all about the potential power of institutions, not to mention one of those at the heart of the decarbonisation debate, is Fatih Birol, Executive Director of the International Energy Agency since 2015. The IEA was founded in 1974 to co-ordinate a collective response to the major disruptions to the oil supply in the 1970s. Its mission statement describes itself 'at the heart of global dialogue on energy, providing authoritative analysis, policy recommendations and real-world solutions to help countries provide secure and sustainable energy for all.' Although initially created to focus on oil supply, today it focuses on all forms of energy, with an increasing focus on clean energy technologies.

Fatih spoke to me from his office in Paris and gave me a clear sense of how energy is a global issue. He explained how the IEA's most recent report had focused on the importance of reforms to both energy generation and consumption, with wind and solar power a top priority. He spoke, too, of his concerns that unless action was taken, then following the end of the pandemic, the world would see a rebound in greenhouse gases.

'In Europe and in parts of the United States, some suggest consumer behaviour is changing. That after COVID, people will become much more environmentally conscious. But even if in Paris, or in New York and London, people do change their behaviour and, say, give up importing pineapples from Costa Rica, at the same time in Bangladesh, in Nigeria and India, tens of millions of people are going to buy their first refrigerators. What will happen with these people, and what sort of energy-using goods are they going to buy?'

Fatih stressed the increasing role of air conditioners in energy use. 'In China, India, Indonesia, the bulk of the emissions are coming mainly from electricity plants, and the number one driver of electricity consumption by far is air conditioners. Today in Asia, two out of ten households have an air conditioner. Whereas in Japan, it is nine out of ten. So these new air conditioners will become a major driver of electricity consumption, which in turn will lead to building more coal plants, more power plants, and emissions will go up. What's more, in places like India air conditioners often need three times more electricity because they are inefficient.'

In terms of climate change, Fatih stressed it made no difference where the energy use was coming from: 'We all know that one tonne of emissions going into the atmosphere from Jakarta or from Paris or from Detroit has the same effect. Emissions don't have a passport. They are all creating the same problem. Therefore, what we do in Paris or in New York has very little meaning if we don't have a global perspective. And when we get the global perspective, the first purchase of durable goods, and their efficiency standard, it is extremely important.'

I asked Fatih about what he thought the best ways forward were to tackle the climate crisis. We discussed the idea of carbon pricing. This is something I am particularly interested in, being a commissioner for the World Bank's Carbon Pricing Leadership Coalition. A 2019 report by the CPLC emphasised both the need to decarbonise our economies and how a widespread use of carbon pricing will be integral to that goal. Well-designed policies can help to shift investment decisions, consumer behaviour and perceptions about competitiveness. The scope for change is clear: at the moment, just 6 per cent of global taxes cover environmental pollution.

Fatih agreed that having some sort of carbon tax was the 'number one solution' but also talked about the difficulties in such a scheme. 'A carbon tax is very efficient, a very good idea. But I'm a realist. The chances of having a carbon tax that everybody agrees to is all but impossible: in one country, yes; in another country, no. So, introducing a global carbon tax, practically, the chances are zero. Some countries will go ahead anyway. Europe is doing it. China will do it for some parts of the country. But globally, the chances are very, very low.'

Instead, Fatih talked about the two key solutions being what he described as efficiency and renewables: 'For me, a big thing is that in India, air conditioner manufacturers now have to obey new regulations. Some countries are coming on board with efficiency, but others aren't. Europe, Canada, China, India, Japan,

Korea, are all pushing, but not everybody is. Another big danger we are going to face over the next few years is that the energy prices will be low. If the energy prices are low, people's incentive to save energy will be less. It will be more difficult to convince governments to take action, but I still believe energy efficiency is the number one thing we can do. I always say that some countries have the potential for solar, some for hydro, some for gas, but all countries have potential to make use of energy efficiency.'

As for the different sources of energy, Fatih talked me through his thoughts on different sources. 'What I see is that coal is in terminal decline, but that oil and gas will be with us for some time to come. If the governments of the world get coronavirus under control and the world economy rebounds, then oil and gas

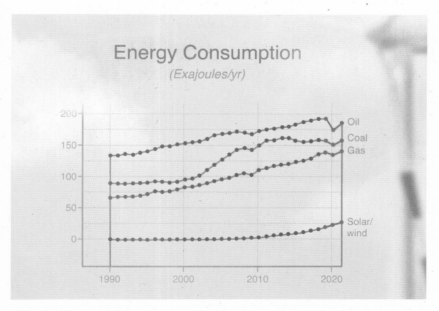

'World Scientists' Warning of a Climate Emergency', *BioScience*.

(Source: William J. Ripple, Christopher Wolf, Thomas M. Newsome, Phoebe Barnard, William R. Moomaw)

demands will also rebound unless governments take action. Look at China: the economy there as of September 2020 had almost rebounded to the levels of 2019, and Chinese oil demand is back to 2019 levels too.'

Fatih explained that one of the challenges with oil production is holding to account the companies that aren't publicly listed. 'Some businesses are doing a good job. Some of them are not. When you open the papers, it's BP does this, Shell does this, but the companies that we read about in the media, BP, Shell, Total and the others, their share in global oil production is less than 10 per cent. The remaining 90 per cent is from companies which are not publicly listed. Nobody dares to deal with those companies, and that is where the bulk of the emissions come from.'

I asked Fatih about other sources of energy such as nuclear: 'On nuclear, we don't have the luxury of excluding it from one of the solutions. Today in the advanced economies of Europe, the US, and Japan, the cheapest electricity comes from existing nuclear power plants. In France, three-quarters of the electricity comes from nuclear power and nuclear is the number one source of zero-emission technology. Nuclear is growing: in five years, China will overtake the United States as the number one nuclear power of the world. But nuclear also has many implications beyond electricity, beyond energy.'

I also wondered what Fatih thought about the development of hydrogen as an energy source, which a number of people are talking up as a solution. 'Everybody loves hydrogen. Why? Because you can produce hydrogen from different fuels: from solar, from nuclear, from coal, from gas. And you can use hydrogen for electricity, for the industrial sector, for trucks, homes, heating, everything. But there are two big issues. One, it is expensive. Second, we don't yet have the infrastructure for it. But many governments are putting huge money into hydrogen, huge money. In my view, hydrogen is today what solar was ten years ago.'

Fatih is very keen to keep all energy options open. 'In general, I would say that the challenge is so big, the scale is so huge, that we don't have the luxury to exclude any of the clean energy technologies. Some countries may not like one type of energy. No problem, look at another one, but overall we need all of them. The main work, though, is going to come from renewables.'

It is here that Fatih sees the heavy lifting happening over the next decade. 'In the energy world in the last thirty years, there's been a motto: "Coal is the king." But when we came out with our most recent IEA report and I looked at the numbers, I said, "The new king is solar." When you look at the next ten years, the biggest contribution to the electricity generation is coming from solar.

'Firstly, it is becoming the cheapest source of electricity. And secondly, it is spreading. Until recently, until 2017, it was mainly a Chinese phenomenon: 50 per cent of all solar came from China. But in the last two years, everybody is there: India, Indonesia, Vietnam, Africa, the Middle East, United States, Europe. So China's share in the new solar is now less than one quarter. And now everybody's doing it.'

• • •

If, as Fatih Birol suggests, the new king is solar, then that feels an optimistic note to end this chapter on. In a deeper way, of course, the real king has always been solar: that's what heats our planet, after all. It's just how human activity over the last two centuries has affected how the planet copes and disperses this energy that has become the issue.

One ongoing discussion among scientists is whether we have reached a new stage in the planet's history. For the last 11,700 years, since the end of the last major ice age, we have been living in an epoch called the Holocene. But such is the extent that man's activity has changed the planet that many argue we have now entered a new unit of geologic time: the Anthropocene. The term, created by biologist Eugene Stoermer and chemist Paul Crutzen in 2000, denotes an era during which, for the first time, human activity is having a significant effect on the planet's climate and ecosystems.

There's little doubt that the way society has evolved since the Industrial Revolution has left a deep scar on the earth's resources. And there's little doubt, too, that the amount of time left to do something about this is rapidly shrinking. But while mankind is responsible for all of this, what I also feel buoyed about is that mankind can get us out of that. Look at an agreement such as the Montreal Protocol: while the situation now poses different challenges, such an achievement shows what is possible in tackling environmental problems when we work together.

I also take great heart and encouragement from the people I have spoken to in writing this book. There are great individuals out there, brilliant, creative minds coming up with ideas and solutions to deal with these challenges. In many cases, we could do much more to support these people, making it easier for their products to come to market, finding a way of helping capital flow more easily in their direction. But the brilliance and innovation to help us through is there.

I also feel encouraged by the sense that all of us can, in our way, make a difference to the environmental cause. For business leaders, there is so much we can all do to change our companies for the better. And as the examples in the book show, it can be done not as a drag anchor on the business, but as a strategy that improves results. And as citizens, too, we can all play our part: small individual changes can add up to making a big difference.

James Lovelock's observations of the haze above the Wiltshire countryside were the initial thoughts of one person. But the snowball effect from that to a global agreement shows just what is possible, when we put our minds to it.

DOWN TO EARTH

North Col on our return was like a ghost town. Lukas and the other guides, together with the majority of the Sherpas, had already headed on down to Advanced Base Camp. All that was left of the expedition was Fred, Marco, Gabriel and myself, together with a few remaining Sherpas. I was grateful to find that the Sherpas had brought down the gear that I'd left at Camp III and hadn't taken it further down. Although my air mattress was in shreds, at least I had my sleeping bag to see me through the night.

Around the fire, Gabriel, Fred and Marco were silent. Nobody spoke, each of us lost in a haze of thought, processing what we'd just been through. I could feel myself beginning to thaw and defrost: my summit suit was frozen inside and out. The moisture from my breathing out had frozen solid on the outside, to the point that it was like wearing a piece of armour. From the heat of the fire, I could feel my clothes becoming damp, then wet, then soaking. I tried to eat some of the soup the Sherpa brought me, but I didn't feel like eating. And even when I did, my hands were shaking too much to hold the spoon properly.

When I was finally able to summon up enough energy, I struggled back to my tent and changed into what remaining dry clothes I had. I then attempted to go to the loo. I was aware that I hadn't peed for thiry-two hours. I found my bottle and after some effort, managed to urinate. I blanched. Looking down at the bottle I was startled by the colour of my pee – it was the colour of red oranges. That couldn't have been a good sign.

All the while, I could hear Lukas arguing over the radio with the Sherpas. He wanted us to move. That argument went back and forth for about half an hour, but the Sherpas knew they were on a losing battle if they tried to shift us. I knew there was no way I could go another step. But even so, hearing the urgency in Lukas's voice worried me.

I was too tired to think straight. Instead, I zipped the tent back up and got into my sleeping bag. Looking back, the fact that I was alone should have been another red flag: on the way up, Lukas had been insistent that we needed to sleep in pairs in case anything happened. By this point, his warning didn't cross my mind. I checked the amount of oxygen I had in my tank. I put it on a low setting and almost immediately fell asleep.

At four in the morning I woke with a jolt, my eyes immediately open. My lips felt numb. Something was seriously wrong. I switched on a torch and looked down to check my oxygen saturation. My saturation level was low, really low. Eighty-five. It was never meant to get less than ninety. I looked over at my oxygen tank. It was empty; it must have run out while I was sleeping.

'Sherpa!' I shouted. 'Sherpa!' My voice was weak and I felt faint. Outside, I could hear the whistle of the wind. Would anyone hear me above it? I waited for the telltale sound of someone shuffling across on loose rock. But all I heard was silence. I shouted again. And again. I kept on shouting, each time feeling the emptiness in my lungs and the thinness of my remaining breath.

In reality it was probably only a few minutes before I heard a scrabbling noise and the clink and clang of an oxygen tank. But lying there, it felt like forever. The tent unzipped and the welcome sight of a Sherpa appeared. Once the tanks were switched I breathed in deep, sucking the fresh oxygen down into my body. I could feel the oxygen pulsing through my body as my saturation levels rose back up. After the Sherpa left, I turned the oxygen levels up a little higher still and drifted back to sleep. Those next few hours were perhaps the most beautiful sleep of my life. When I woke again at about 7.30, I felt refreshed, a completely different person.

Crawling out my tent, I saw Fred standing there, bleary-eyed. I told him what happened. Fred laughed. 'I didn't even wear my mask last night.'

'Why on earth not?' I asked. 'That's so dangerous. You could have been killed.'

'I find it a bit uncomfortable to sleep in,' Fred said, uncomfortably. 'Still, I'm OK, aren't I?'

A few tents down, I could see a small group gathering around the entrance. As one of the other climbers passed, I asked what was going on.

'Someone's died,' he said. 'An Irish guy. Do you know him?'

I realised with a gulp that I did. This was the guy who I'd noticed was out of it on the way up, who I'd tried to stop from summiting. And now he'd gone to bed and woken up dead. I looked across at Fred: there but for the grace of God, I thought again.

North Col, it turned out, was a ghost town in more ways than one. The sooner we got out of there, the better.

• • •

The weather, at least, was better than the day before. The sun was now shining and the wind had dropped. The mountainside should have been crystal clear in all that sunlight, but to me, it all looked a little cloudy. When I mentioned this to someone and they looked at me blankly, I got a bit concerned. The fog, I realised, wasn't like normal fog, but seemed to start straight in front of my eyes. And I could see things further away, just not in much detail. Was this the onset of some sort of snow blindness? I shook my head and double-blinked, in the hope that it would just go away.

The first challenge that day was to make it back down to Advanced Base Camp, where Lukas and the other guides were already waiting. This was the final section of steep descent, including retracing our steps back down the 'Wall'. As we prepared to leave, I felt my body shaking with The Fear. I realised I was really nervous – to be so close to safety made it feel as though the stakes were ramping up. This is the last leg, a voice rumbled inside my head. This is where it goes wrong and you can't get back.

The memories of the previous days pulsed through me. How many lives had I used up getting back down to North Col? And how many did I have left? The death of the Irishman a few tents down had shaken me more than I cared to admit – a reminder that although we might now be below the death zone, death was still a stone's throw away.

I was now without the Sherpas who'd been there to guide me up and down. I'd been assigned a couple from among those who were left. I didn't know them, and they didn't look pleased to have picked me. They looked even less pleased when I gave them a pep talk.

'I've got to admit it,' I said. 'I'm going to need your help here. I'm feeling quite weak. And I'm feeling quite scared. I'm really worried about falling. I need one of you in front of me and one of you behind, the guy in front to arrest my fall if I go down. I want the guy behind to be there to clip onto when we go off the lines.'

The two Sherpas looked at each other as if to say, what have we ended up with here? But it was as good as I could do. The descent back down to Advanced Base Camp should have been quick, but rather than taking me thirty, forty minutes, it took me over two hours, well behind everybody else. I was nervous with every footstep that I was about to slide over. My feet were really starting to ache. And my eyes were increasingly a worry. Not being able to see straight as I walked was very unnerving.

On the way back from the summit.

Somehow, I'm not really sure how, I made it back down the Wall. I was relieved never to have to confront that again. And then there was that priceless moment, further down, where I reached the edge of the snowfield. Suddenly there was that ledge, and it was time for me to take off my crampons. Part of me was flooded with relief at that, but I still had this voice in my head.

It's not over yet, it whispered. You're still not safe.

I carried on walking down gingerly. Without the crampons, the pain in my feet seemed to increase. It was strange – on the way up, I'd had to learn how to walk with crampons on. Now I was readjusting to having to take them off again. I was incredibly conscious of the possibility of falling and slipping, of twisting an ankle and not being able to make it any further. The Sherpas were keen to push on but one breath, one step, was as fast as I could go.

Finally, ahead of me, I could see the coloured canopies of Advanced Base Camp. It was blurry through my eyes, but even I could see the colours for what they were. The closer I got, the more emotional I felt. I could feel the pain of each footstep; I was desperate for the camp to come closer into view, until, finally, I was there, and I collapsed into a tent with exhaustion.

• • •

Lukas was, in every sense, a sight for sore eyes.

'I can't see,' I told him, or at the least the fuzzy approximation of him I could see before me. 'Everything is a bit blurred.'

'It's OK,' Lukas said, leaning forward to inspect me. 'I'm going to give you some drops for that. Does anything else hurt?'

'My feet,' I said. 'They're killing me.'

'OK,' Lukas said. 'We'll have a look at those as well.'

The drops stung my eyes and caused them to water up, but almost immediately I could notice the difference. Everything was still a little hazy, but the fog had lifted.

My feet were a mess. I needed to get out of my summit boots anyway at this point, and switch to another pair of boots to walk back down to Base Camp. This was pure agony. My feet looked terrible – all the skin had gone. Lukas took one look at them and said, 'I'm going to give you some painkillers.' I made to argue, but then decided against it. The painkillers were super strong, the sort that you're not meant to drive after taking. As I pulled a pair of fresh socks on, I winced in pain, then howled as I squeezed into my new boots. But within a few minutes, the drugs had kicked in. Not only had the pain gone but so, it seemed, had my feet. They were now so numb, I couldn't really feel them at all. Whereas before every step was painful, now I couldn't even sense when my feet were on or off the ground.

By now, I was hungry, really hungry. And at Advanced Base camp there was, finally, proper food to eat. I ate bread. I ate daal. I ate jam, cheese, dried meats, Nutella. I had a load of soup and hot mugs of tea, and a weird-tasting orange drink that helped with dehydration. Everything that was laid out for us on that table I ate. So much so, and so little had I eaten over the previous few days, that I immediately felt uncomfortable.

Fred and Marco didn't hang around. They wanted to get back to Base Camp as quickly as possible, so had something to eat and pushed on. I couldn't do that, and I couldn't go down at the speed they were moving.

'Will you stay with me?' I asked Gabriel. I knew that from Advanced Base Camp downwards, I'd no longer have any help from the Sherpas and didn't want to walk on by myself. 'I need someone to look after me, if that is OK?'

Gabriel was a true friend. 'Of course,' he replied. 'Whatever you need.'

• • •

What I needed, it turned out, was company. The distance from Advanced Base Camp back to Base Camp was about twenty-five miles: leaving Advanced Base Camp at about 11.30, it was going to take us until dark to get back.

Lukas's drugs had sorted my feet and my eyes, but there was little they could do in terms of how I was feeling inside. The further Gabriel and I walked on in that final stretch, the more emotional I got. I thought of Stephanie, I thought of my children, and everything that I'd been through in the last few days hit me in a rush. Before I knew it, I was in a flood of tears, walking along with drops streaming down my face. My nose was running, my eyes were running. I was glad there weren't many people around to see me like this.

What I kept thinking wasn't what I'd achieved, but how stupid I'd been for attempting anything like this. How close had I come to dying? And for what? How much of this trip had been about the environment, and how much had it been about my ego? It was one of those moments where the cold wind of truth cuts through. What had I been thinking? What would Stephanie and the kids have done if I hadn't made it back?

The words of a poem I liked, 'The Man in the Glass' by Peter Dale Wimbrow, echoed through my head. 'You may fool the world down the pathway of years,' the poem goes, 'and get pats on the back as you pass, But your final reward will be heartache and tears, If you've cheated the man in the glass.' Is this what I've been doing? I asked myself, climbing Everest just to get pats on the back? Who even was the man in the glass I was staring at?

I promised myself, as the tears continued to stream down, that I would change. From now on, I swore to myself, I would listen to Stephanie, my beautiful wife. I would listen, and I would hear, really *hear*, what she was saying. I would be better with my children as well, be a proper part of their life. That I was here and Stephanie was there, looking after them, showed how much the Everest journey was just a microcosm of how my life was structured. That had to change. I needed to be a real father, and be more actively involved in bringing them up. The more I thought about my family, the worse I felt about making this trip in the first place. If I'd died on that mountain, there would have been three messed-up children I'd left behind.

My third realisation on that walk down was just how little I'd been aware of what the body can do. That near-death experience had showed me first-hand just what the body and mind were really capable of. Now I knew, I wanted to explore more and make the most of it. The way I had been living my life, I had only been scratching the surface of what I could achieve. At the same time I thought,

On the path to Base Camp.

too, about how little time we all have on this planet. I felt blessed to have been given this moment, and having come back alive, I was determined not to waste any second of it. From now on, whatever time I had left, I was going to make sure that I made the most of it.

As we continued to descend through the amazing landscape, the shadow of Everest behind us and sweep of the valley before us, I had that realisation again that I'd had at the summit: how small we all are in the face of nature. We might feel like we're kings of the world, here to make our mark, but really, we're specks of dust against the strength and might of the natural world. Who are we, as a species, to try and fight nature, to cause all the damage we do to it? That could only come from ignorance and a lack of understanding. In that moment, I felt I properly understood this for the first time. And in doing so, my sense of ego disappeared. Against nature, my individual actions were immaterial.

The experience of climbing Everest had changed me in a way I hadn't expected. I felt as though I had gone up a boy, and come back down a man. The experience had stripped me, stretched me, and reshaped me into someone else. Climbing to the top of the highest mountain on earth was something, but what I'd learned in the process, I realised, was much more valuable.

The light was fading by the time Gabriel and I made it back to Base Camp. Everyone else was already there. There were attempts at a proper celebration, but to be honest, we were all just too exhausted. The following day, it would feel different, but that night the overriding feeling was summed up by Fred falling asleep, head first in the cake that had been prepared for him.

I didn't feel much like celebrating. Instead, I left everyone to it, and slipped out to find a signal and speak to Stephanie. I waited for the dial tone to ring and then, for the first time in days, I heard her voice.

'Hakan?' her voice crackled. 'Is that you? Are you OK?'

'I'm safe,' I said, struggling to hold back the tears. At that moment, I felt overwhelmed by gratefulness. Grateful to have Stephanie, grateful to have my family, grateful to Mother Earth for having let me get through the last few days.

'I'm coming home,' I told her.

● ● ●

When I made the decision to climb Everest, I did so for a number of reasons. As well as raising awareness about the environment, I wanted to use the attempt as a springboard for change within the company. I wanted to be able to look my children in the eye and tell them what I was doing myself. And I wanted the personal challenge – to stretch myself and see what I could do.

Three years on, and my life has changed.

My relationship with my wife and my children has never been better. Whereas once I was often away for business, now making time for my family is a priority. And not just time, but quality time. If my children want something, I'll stop and find the space for them. My office door at home used to be permanently closed; now it is always open for them. I continue to share and grow their relationship with and passion for nature. Thanks to Sasha, meat-free meals are now a regular part of my diet.

At work, the success of my expedition has led to sustained and real change in how we do business. Both within the company and across the wider industry, my Everest experience is a way in to people thinking about and taking the climate seriously. Before I went, I knew that whenever I did a talk or a presentation, even on a subject as serious as climate change, within five minutes people would be glazing over or checking their phones. But since I've been back, whenever I've started talking about my climb up Everest, everyone puts their phones down and starts to listen. Everest, I realised, was my way of getting across what I wanted to say. Rather than people going through the motions when it comes to the environment, now they take the issue seriously. Within Arçelik, it has changed the dial about sustainability. It has helped to refresh the company, and give us an identity and ideal that we can all be proud of.

It's helped that I've changed my leadership style too. I listen a lot more than I used to. I trust people a lot more as well. I think the people around me have become better at what they do because I've given them more space to do their jobs. I believe now more than ever in the importance of subject matter experts: if an area is your expertise, then I'm going to value your opinions on it, rather than think I know better. The only time I weigh in now is when there are competing interests involved and a decision between them needs to be made.

My time is now a lot more precious. Take my schedule: nobody can waste my time any more. I don't have the space for people coming in to tell me how brilliantly they're doing. Long meetings and presentations in the company are a thing of the past. When I first came back, I tried a reset of five-minute videos and fifteen-minute slots for meetings. That has relaxed a little since then, but my schedule is still carved up into twenty-five and fifty-minute chunks: I really believe that there are very few subjects that you can't tackle in a short space of time if you are dealing with a subject matter expert who already knows the facts.

I would like to think I've become much more sensitive to the people I work with. I'm more aware now that I'm dealing with people who have feelings, families, their own dreams and aspirations. It's very easy in business to use the stick as a motivational tool, but I no longer believe that as a philosophy. People are fundamentally good, and they want to do good. And that's how I approach my team and the people I do business with.

But perhaps the biggest change from climbing Everest for me has been the personal one. When I set out, part of the appeal of the expedition was the challenge, but I saw this challenge in physical terms. I hadn't anticipated how much it might change me in other ways.

When I got back to Istanbul from Tibet, my communications team wanted to make a big campaign out of my trip to Everest. I remember Fred saying on the way home, 'You're going to be a hero in your country. Only a handful of people have done this before.' Before I went, I would have jumped at the chance to do that. But I was a different person now. Rather than shouting from the rooftops about what I'd achieved, I'd found the experience so humbling that by the end I didn't want to do any of that. It felt wrong, somehow. I needed to go away and process everything, and let it all sink in.

Everest, it turned out, was a mountain to climb, just not the one I was expecting. The experience changed me in ways I hadn't considered. And the fact I hadn't considered how it might change me spoke volumes about the person I was. I'd set out with the arrogant belief that I had the power to save the planet. Instead, the planet and the power of Mother Nature had helped saved me from myself. There's a line from the old Joni Mitchell song about not knowing what you've got until it's gone. That's true about the environment, but it was true about what mattered in my individual life as well: I wouldn't put that – and them – at risk in the same way ever again.

My ascent of Everest taught me about the capacity of all of us to do things differently, and the untapped potential we all have within us. That's true in our personal lives, and it's true in terms of the environment as well. Humans, as I said in the last chapter, are responsible for the state of the environment, but humans, together, also have the power to do something about it.

Yes, tackling climate change is one hell of a mountain to climb, but as my experience has taught me, even the tallest peak is possible to overcome.

GOOD COP? BAD COP?

In November 2021, I flew to Glasgow to attend the United Nations Climate Change Conference – or COP26, as the conference was more commonly known. I wanted to witness for myself the mood and feel of the world's governments and companies on the issue, and to add my own voice to the need for change.

Glasgow in November is not normally at the top of the list of travel destinations, and the looming grey skies when I landed did little to dispel an atmosphere of uncertainty and tension surrounding the conference. With the world descending on the Scottish city, finding somewhere to stay was no easy task. As a businessman, I appreciated the opportunism in local Airbnb hiring out apartments at extortionate prices: even if my credit card didn't quite share in the appreciation! I booked an apartment overlooking the conference centre, and arrived on Sunday, ready for the main events to begin.

On the Monday morning, I opened my curtains to see that the huge COP26 sign had been torn down by the wind during the night. That's not a promising metaphor, I thought. Further down, off a bridge over the water, were hanging two protesters, winched down and holding their own sign, that read 'Humanity is Failing.' I later learned how young these protesters were – twelve and ten. I don't know if it was a deliberate echo, but like humanity's chances, it felt as though they were hanging there by a thread.

OK, I thought. This is going to be interesting. As I walked to the conference centre, there were protesters everywhere. I tried to engage with them, but they didn't know who you were and as soon as they saw you had a conference pass, immediately felt that you were part of the problem, rather than a solution. But not all of the protesters were like that, I must stress. While there, I went to a talk at the *New York Times* pavilion, and was completely blown away by a young activist called Clover Hogan. Clover is just twenty-two, a researcher on eco-anxiety and the founding executive director of a non-profit organisation, Force of Nature. Her knowledge and understanding of the situation was astounding. She spoke clearly, eloquently and shockingly on the disenfranchisement of young people in the debate. Hearing people like her speak, it felt like the next generation might be in safe hands.

Once inside the venue, COP was divided into two zones: the blue zone, where negotiations between governments and world leaders were taking place; and the green zone, which was more like a traditional exhibition with stands showing environmental innovations around the climate and talks and discussions. The size of the government delegations seemed way too large to me.

There were endless photographs of delegates taking pictures of themselves underneath a huge hanging globe and posting those on social media, rather than actually doing anything. For the next COP, I think these delegations could be slimmed down substantially, just bringing those who are actually involved in the negotiations.

Next time, too, the world leaders should be scheduled to attend at the close, rather than the beginning of the conference. Having them there at the start, making speeches and leaving, meant that they weren't around to influence the business end of proceedings. What happened was that the negotiating teams were working through the night, but then had to phone back to their governments to discuss. It would have made much more sense for the leaders to be around at this point.

In the blue zone, the various pavilions felt in sharp contrast to each other. On one side, there were those for different environmental groups and those at the sharp end of the climate crisis. One stall I was struck by, particularly given my expedition up Everest, was about the effects of temperature increases on glaciers. Opposite these were stands representing different countries – Qatar, Saudi Arabia and Australia – plying people with expensive coffee and showcasing how amazing they are. One joke at the conference was that the only thing Australia brought to the table was good coffee! It was bizarre. Arçelik didn't have a stall anywhere in the conference area: it didn't even cross my mind that we should have one, though other companies were clearly using the occasion to talk up their green credentials.

Instead, I used my time to meet and to listen to people: to speak on different panels about my experiences and to attend others, to hear what other people were doing. I did one panel with the Turkish Business Association and another on energy efficiency with ministers from India and Nigeria, and the head of the IEA. I spoke at a panel for Leaders on Purpose – I talked about how my environmental transformation had helped to create a purposeful business and one whose principles helped attract the best talent. I also joined the World Economic Forum's CEO Roundtable, where around ninety CEOs signed a joint commitment in the fight against climate change. I mention all of this, not to try and impress, but to give a flavour of what COP looked like from the inside. And what I hope you'll take from this, as I did, is the number of high-level business people and government ministers coming together, sharing ideas and discussing ways forward. That felt exciting and full of possibility.

One event I was particularly proud to take part in was regarding the Terra Carta. Created by HRH the Prince of Wales, this is a charter that aims to put sustainability at the heart of the private sector. It was launched at the One Planet Summit in January 2021, setting out a roadmap over the next decade for businesses to help the planet: in the same way that the historic Magna Carta inspired a belief in rights and liberties, so the Terra Carta aims to do the same for nature.

At Glasgow, Prince Charles announced the Terra Carta Seal – awarding it to 45 organisations that are demonstrating their commitment to genuinely sustainable markets and driving innovation and leadership within their industry. I was delighted and honoured that Arçelik was one of the companies chosen for this award: I was doubly proud that we were the only company from our industry awarded, and one of the few from emerging markets. The event took place at Kelvingrove Art Museum – a beautiful building, complete with vaulted ceilings and ornate chandeliers. It felt very fitting. I had the chance to talk properly to Prince Charles and was hugely impressed. His knowledge on climate change and commitment to the cause were clear.

The Terra Carta Seal ceremony wasn't the only event that I went to at Kelvingrove: I was back there to listen to another panel on sustainable fashion, with leading designer Stella McCartney. As I've described elsewhere, we've worked hard to develop washing machine filters to counter the microfibres of much fast fashion. As I was watching the discussion, I realised who I was sat next to. Leonardo DiCaprio! DiCaprio created his own foundation to foster sustainability back in 1998. It was great to see him there supporting the cause.

• • •

In terms of the discussions themselves, how successful was COP as a summit? The first thing to say is that you were never going to deal with decades of environmental damage in just ten days. I'm realistic about that, and also understand that there is a lot to digest in terms of what was agreed, and that some of those decisions will take time to bear fruit. But in that context, I'd say that my conclusion on the conference remained one of hope laced with caution: it left some things very much in the balance, but the possibility of success still there, just about.

There were lots of positives to take away. The first of these was over NDCs – the Nationally Determined Contributions that each country agrees to make to combat climate change. In the past, these were made every five years. Now, these will need to be made annually, with the International Energy Agency, and my friend Fatih Birol given the role of gendarme to police that. That feels a far stronger commitment and one that is more likely to be properly enforced.

Secondly, there was the agreement over a framework for what is known as Article 6: this was an unfinished part of the Paris Agreement that dealt with the subject of carbon markets. Essentially this is a set-up where one country or company can buy carbon credits from another. So if country B has reduced its carbon emissions by more than it promised in its NDC, it can then trade these carbon credits with country A, so that they too can achieve their NDC. There is an incentive for countries to go beyond the emissions cuts agreed, as they will be financially rewarded. It's a great idea in theory. In practice, it has been much harder to get off the ground. Finally, five years after Paris, moves towards a proper infrastructure to such markets was agreed – with sticky issues such as double counting and legacy credits resolved.

The result of this agreement is that the carbon markets are going to grow dramatically. It's also going to create a more internationally aligned price on carbon: at the moment, there are discrepancies between countries which, once properly tradeable, will quickly balance out. In terms of business and how companies behave, I think you're going to see firms starting to voluntarily offset their carbon way earlier. I'm finding it very promising that corporates are beginning to understand that they're going to get caught on the wrong side of this if they don't act fast. And where business leads, I think you'll find that countries will follow – there'll be pressure to make sure these markets work properly and efficiently.

We have already seen unprecedented increases in the carbon prices in the regulated markets. Take the EU ETS (the European Union's Emissions Trading System), for example. During November the price was around €67 per tonne of carbon; in a month, it had risen to almost €90. The increase in these regulated markets also implies an increase in the voluntary markets as well.

Companies are beginning to see that conservation efforts are not enough and the only way to achieve net zero targets is one where the carbon sequestration plays an important role. Carbon sequestration happens in two ways: nature-based solutions, such as afforestation and reforestation projects; and direct air carbon capture and storage technologies. As the latter is in very early stages of development, nature-based solutions, especially blue carbon projects, seem to be the most viable option in the short term. A good example of this are mangrove forests in Raja Ampat. If you can reforest a mangrove forest and maintain that going forward, the amount of carbon it contains is huge – mangroves soak up to thirty times the levels of carbon that a normal tree does. In the years ahead, funding for such projects, because of the carbon credits involved, will be much easier to come by.

The role of the financial markets was also an important achievement of COP26. The financial markets are beginning to understand that they are going through a period of transformation. Since the Paris Agreement, the financial industry has made more than $17 billion in fees from facilitating $4 trillion of fossil-fuel financing. But by 2036, recent analysis suggests that about half of the world's fossil fuel assets will be worthless, with about $11 trillion left in stranded assets. There might have been money to be made in financing fossil fuels in the past, but now the smart money is on moving away. At COP, 450 financial institutions from over 45 countries agreed to commit to net zero. That's about 40 per cent of global financial assets (around $130 trillion and compared to $5 trillion committed in 2019). Committing to net zero isn't the same as investing in climate initiatives, but it's a definite step in the right direction.

Two of the biggest announcements at the conference involved India. At the start of COP26, India announced that they would commit to net zero by 2070. And at the end of the conference, they were one of the main drivers behind a watering down of the final communique – agreeing to 'phase down', rather than 'phase out' the use of coal. Those two announcements might seem a bit contradictory, but actually make more sense when you think about it.

The Indian commitment to net zero was both surprising and welcomed. It wasn't 2050 as was one of the original goals of COP26, but the significance of the commitment should not be underestimated: India is the fourth biggest emitter of carbon dioxide in the world. At a stroke, its announcement reduced the global temperature rises predicted by 2100.

How a country such as India achieves this aim, especially with the predicted growth of its economy, is incredibly hard. In Mumbai, they're hitting temperatures of 50 degrees on a regular basis. The role of air conditioning, as the economy matures, is something we've discussed in a previous chapter. There is a growing energy need there that will still need to be dealt with.

One way of helping with this is via better energy efficiency. Efficiency is described by the IEA as 'the very first fuel' in helping achieve decarbonisation. There are various possible measures to help cut down on emissions – one of which is raising energy efficiency labels and increasing product efficiencies across the globe. Four appliances used in our homes (refrigerators, ACs, electric motors and lighting) add up to 40 per cent of all global electricity consumption. Doubling the efficiency of these four segments, whether by innovation or regulation, could reduce energy consumption by 6,700 terawatt-hours per year globally – equivalent to the generation of more than 3,000 medium-sized coal-fired power plants, and avoiding 2.9 Gt of CO_2 emissions per year.

The changing of the wording from 'phasing out' to 'phasing down' felt disappointing. For some of the delegates it ended the conference on a low note. Sure, it would have been great to get every country to commit to phasing out coal. But for countries like India and China, they can't just switch over straight away. When China stopped buying coal in 2021, energy prices quadrupled around the world. They had to shut factories down – supply chain inefficiencies started appearing. Moving to different energy sources requires a period of transition – and finance. One of the exciting announcements at COP was that South Africa had received $8.5 billion in funding from the US, EU and UK to help it transition away from its traditional use of coal to more renewable energy sources. If this plan works, it could serve as a template for other countries to follow.

I'm actually more positive about this wording than some people. It might look like these countries are giving themselves an out, but my guess is you'll see them agreeing to phasing out in the next few years, once it is clearer as to how to do it.

There were promising attempts for a green transformation both on the public and private side: during the conference, six big car makers – including Ford, Mercedes-Benz, General Motors and Volvo – and thirty governments pledged to phase out the sale of gas and diesel-powered vehicles by 2040. On the ground, I was particularly taken by Rolls-Royce's plans for a hydrogen jet engine, with the aim of making flying cleaner.

There is obviously still a huge amount of work to be done. At the end of the conference, COP president, Alok Sharma – who had earlier incidentally depicted the course of the talks as 'a mountain to climb!' despite progress – was in tears and apologised. Depending on which analysis you read, the commitments at COP26 mean a rise in global temperatures of between 1.8 and 2.4 degrees by 2100: a definite improvement on the 2.7-degree increase predicted under previous pledges, but still not enough to maximise the increase at 1.5 degrees. That figure is still alive – just – but requires further commitments to get there.

And there remain potential pitfalls ahead that could take things the other way. The biggest threat probably lies in the domestic politics of the United States. It's too early to know what might happen, or even if he will run again, but the re-election of Donald Trump in 2024 could well see the reversal of some of these gains. If he had been in power during COP26, those negotiations might well have turned out differently. But that said, for all his previous pronouncements on climate change, Trump remains at heart a businessman. If he saw there was money to be made here, I wouldn't put it past him to do a volte face.

Maybe one way forward with all of this is not to think about the situation in terms of fifty or a hundred years ahead. Once you think of issues in these terms, it is easy to put decisions off. But breaking time into shorter units can make it easier to initiate change. It's something we've tried at Arçelik – instead of describing planning commitments by 2030, we've looked at it by thinking, this is (at the time of writing) 417 weeks away. That gives us a different way to manage the time: in terms of the environment, what have we done this week?

There might be something in this for all of us. As leaders, business people, as individuals, we can all ask ourselves what we've done to help the planet each week. Tackling climate change, as I've argued, is a mountain to climb. But one way of thinking about climbing a mountain is to do so step by step – and believe me, I know. If we can all attempt to make a difference, one step, one week at a time, who knows what we might be able to achieve together?

ACKNOWLEDGEMENTS

I owe gratitude and thanks to many people who supported me throughout this journey that began on the shores of the Aegean Sea and ended on the summit of Mount Everest.

The academics, experts, business leaders, scientists and friends that I've met as I wrote this book have all uniquely contributed to the enrichment of this story.

I want to offer my thanks to Evren, who inspired and supported me through every stage of writing and publishing the book; to Dilara, for her valuable contributions to the preparations ahead of publishing; to Çise, Melina and Savaş, my colleagues who helped me work through every problem; to Yael, Rayka and Nazlı, for their valuable suggestions to the design and for their friendship; to Vera, Nina and Mey, the strong women in my life who have always inspired me, and who have read the earlier drafts of this book and provided insightful feedback.

To Stephanie, my partner in life, who encouraged me, stood by me and became my rock as I moved to make another of my dreams come true.

To my father, whose intelligence I've always admired, and who has stood by me at every critical juncture of my life.

And to all my friends, who showed care and support with their feedback and their insights.

FURTHER READING

Confessions of a Radical Industrialist by Ray Anderson (Random House, 2011)

The Lost Explorer by Conrad Anker and David Roberts (Robinson, 2000)

A Life on Our Planet by David Attenborough (Ebury, 2020)

The Best of Times, the Worst of Times by Paul Behrens (Indigo, 2020)

Our Biggest Experiment by Alice Bell (Bloomsbury, 2021)

Swallow This by Joanna Blythman (Fourth Estate, 2015)

The Climb by Anatoli Boukreev (Pan, 2018)

Making Climate Policy Work by Danny Cullenward and David G. Victor (Polity, 2020)

Silent Spring by Rachel Carson (Penguin Classics, 2000)

Into the Silence by Wade Davis (Vintage, 2012)

The Future We Choose by Christiana Figueres and Tom Rivett Carnac (Manilla, 2021)

How to Avoid a Climate Disaster by Bill Gates (Allen Lane, 2021)

The Wildest Dream by Peter and Leni Gillman (Headline, 2001)

Why Rebel by Jay Griffiths (Penguin, 2021)

Last Hours on Everest by Graham Hoyland (William Collins, 2014)

Into Thin Air by Jon Krakauer (Pan, 2011)

Gaia by James Lovelock (Oxford University Press, 2016)

Homage to Gaia by James Lovelock (Souvenir Press, 2019)

Farmageddon by Philip Lymbery (Bloomsbury, 2015)

Climbing Everest: The Complete Writings of George Mallory (Gibson Square, 2014)

Something New Under the Sun by J.R. McNeill (Norton, 2001)

Net Positive: How Courageous Companies Thrive by Giving More Than They Take by Paul Polman and Andrew S. Winston (Harvard Business Review Press, 2021)

Beyond Possible by Nimsdai Purja (Hodder and Stoughton, 2020)

Doughnut Economics by Kate Raworth (Random House, 2018)

The Deep by Alex Rogers (Wildfire, 2019)

Turning the Tide on Plastic by Lucy Siegle (Trapeze, 2018)

Termination Shock: A Novel by Neal Stephenson (William Morrow, 2021)

Waste by Tristram Stuart (Penguin, 2019)

The Third Pole by Mark Synott (Headline, 2021)

The Uninhabitable Earth by David Wallace-Wells (Penguin, 2019)

Left for Dead by Beck Weathers (Sphere, 2001)

RECOMMENDED DOCUMENTARIES

14 Peaks (Torquil Jones, 2021)

A Plastic Ocean (Craig Leeson, 2016)

Lost on Everest (Renan Öztürk, 2020)

Seaspiracy (Ali Tabrizi, 2021)